10 0662079 X

KU-279-882

Phylogenetic Trees
Made Easy

UNIVERSITY LIBRARY

3 0 JUN 2014

IG

Phylogenetic Trees Made Easy

A HOW-TO MANUAL

FOURTH EDITION

Barry G. Hall

University of Rochester, Emeritus
and
Bellingham Research Institute

GEORGE GREEN LIBRARY OF
SCIENCE AND ENGINEERING

Sinauer Associates, Inc. Publishers
Sunderland, Massachusetts U.S.A.

PHYLOGENETIC TREES MADE EASY: A How-To Manual, 4th Edition

Copyright © 2011 by Sinauer Associates, Inc. All Rights Reserved.
For information address Sinauer Associates, 23 Plumtree Road, Sunderland, MA
01375 U.S.A.

FAX: 413-549-1118
orders@sinauer.com
publish@sinauer.com
www.sinauer.com

Downloadable files to be used with this text are available at
http://www.sites.sinauer.com/hall4E/

100662079X

Notice of Liability
Due precaution has been taken in the preparation of this book. However, information
and instructions described herein are distributed on an "As Is" basis, without war-
ranty. Neither the author nor Sinauer Associates, Inc. shall have any liability to any
person or entity with respect to any loss or damage caused or alleged to be caused,
directly or indirectly, by the instructions contained in this book or by the computer
software and hardware products described.

Notice of Trademarks
Throughout this book trademark names have been used and depicted, including
but not limited to Macintosh, Mac, Windows, and Adobe. In lieu of appending the
trademark symbol to each occurrence, the author and publisher state here that these
trademarked product names are used in an editorial fashion, to the benefit of the
trademark owners, and with no intent to infringe upon the trademarks.

To Sue and Melinda

Acknowledgments

This book could not have been written without the active cooperation and support of the authors of the many programs described herein. In particular, during the course of writing this book major new releases of MEGA, Dendroscope, GUIDANCE, and MrBayes were in progress. The authors of all of those programs made pre-release versions available to me, explained new features, and often incorporated new features and modifications that I suggested. For their help and cooperation I thank Sudhir Kumar and Daniel Peterson of MEGA; Daniel Huson and Celine Scornavacca of Dendroscope; and Tal Pupko, Haim Ashkenazy, Eyal Privman, and Osnat Penn of GUIDANCE.

The new chapter on phylogenetic networks could not have been written without the book *Phylogenetic Networks* (Huson et al. 2011), which the authors made available to me in manuscript prior to its publication. I am grateful to both Daniel Huson and Celine Scornavacca for helping me struggle through some of the concepts in that important book.

Through four editions of *Phylogenetic Trees Made Easy,* the folks at Sinauer Associates have been a pleasure to work with. The graciousness of Andy Sinauer, Chris Small, Jeff Johnson, Marie Scavotto, Dean Scudder, Kathaleen Emerson, and Nate Nolet has been both consistent and remarkable. I am particularly grateful to Carol Wigg, my editor, who has very frequently smoothed out my prose to my benefit and much to the benefit of the reader. She is always a pleasure to work with and I am grateful to Andy Sinauer for making her services available to me once again.

It is impossible to say enough about my wife, Sue Hall. Her encouragement and support are invaluable, as they have been for the last 47 years. Many of her suggestions— little things I never would have thought of—have improved this book throughout.

Finally, to Allan Fox who took me into his lab as an undergraduate; to Jim Crow who showed me the beauty of genetics; to Jon Gallant, Herschel Roman, and Larry Sandler who guided me through graduate school; and to my colleagues Larry Hightower, Linda Strausbaugh, Tony Dean, Dan Hartl, Rosie Redfield, John Huelsenbeck, Miriam Barlow, and Steve Salipante, all of whom strongly influenced my scientific development, a heartfelt thank you for enriching my life.

Table of Contents

CHAPTER 1

Read Me First!

Phylogenetic analysis was once a tool used only by taxonomists, or as they are now called, systematists. Their interests were first in classification, then in elucidation of the historical relationships among organisms. Phylogenetics was considered an arcane and difficult aspect of biology that, fortunately, wasn't widely used by molecular and biochemically oriented biologists. With the advent of DNA sequencing that began to change, and the power of phylogenetics as a tool for understanding biology at all levels became apparent. Today there are few biological journals in which at least some of the published papers do not include phylogenetic trees.

Phylogenetic analysis lies at the core of the new fields of genomics and bioinformatics. The emergence of bioinformatics as an important field, however, has had little effect on the perception of phylogenetic analysis as difficult and intimidating. Those who specialize in systematics, phylogenetics, genomics, and bioinformatics have done little to set aside the sense that phylogenetic analysis is a tool best used by specialists, with the result that molecular and cell biologists often seek collaboration with systematists when one of their papers requires a tree. But in most cases, preparing a robust, valid phylogenetic tree for a paper is no more challenging than using word processing software to write the paper or graphics software to prepare the figures. The purpose of this book is to make phylogenetic analysis accessible to all biologists.

Phylogenetic Trees Made Easy is a "cookbook" intended to aid beginners in creating phylogenetic trees from protein or nucleic acid sequence data. It assumes basic familiarity with personal computers and with accessing the World Wide Web using browsers such as Firefox or Internet Explorer. I have not attempted to explore all the alternative approaches that might be used, intending only to give the beginner an approach that will work well most of the time and is easy to carry out. I hope the book will also serve the investigator who has a modest familiarity with phylogenetic tree construction but needs to address some aspects and problem areas in more depth.

This book is not intended to be used as the primary text in systematics or phylogenetics courses. It can, however, be used to supplement the primary text and can serve as a tool for making the transition between a theoretical understanding of phylogenetics and a practical application of the methodology.

New and Improved Software

The four years that have elapsed since I wrote the Third Edition have seen significant changes in MEGA, the software for phylogenetic analysis and tree construction that is at the core of this book. MEGA has changed significantly both in terms of increased capabilities and in terms of its interface. Details of using some other programs, particularly MrBayes, have also changed significantly. In general the changes have been positive and especially beneficial to those who are new to phylogenetics.

It is neither possible nor desirable to cover all of the programs available for phylogenetic analysis; that would require a tome too heavy to lift and too boring to read. The object here is to describe programs that are sufficient to the task and relatively easy to use. Because I focus more on phylogenetic analysis and less on learning the vagaries of different programs, I try to minimize the number of programs described, and, as time goes on, my perception of the most appropriate programs changes.

The First Edition was written primarily for Macintosh users. Prompted by justified complaints, the Second Edition made a serious effort to ensure that all programs discussed were available for Macintosh, Windows, and Unix operating systems. The First and Second editions focused on using PAUP* for the majority of analyses. By 2006, however, PAUP* had not been updated in several years and the graphical interface version would not run on then-current operating systems.

The Third Edition introduced MEGA 4 as the primary phylogenetics tool, replacing PAUP*. MEGA 4, however, ran only on the Windows operating system; Macintosh and Unix users had to use programs such as Parallels or Wine to circumvent the Windows limitation. The Third Edition also introduced the PHYML program for estimating trees by Maximum Likelihood, since at that time MEGA did not offer ML analysis.

This Fourth Edition introduces MEGA 5, which is now available for both Windows and Macintosh operating systems and will shortly be available for Linux. MEGA 5, however, is much more than a port to different platforms. Because MEGA 5 offers ML analysis, PHYML has been dropped from the Fourth Edition—not because PHYML is inadequate, but because it is just simpler to learn ML analysis within the context of the familiar MEGA environment. MEGA 5 offers many new features, some of which just make life easier for users, but many of which extend its capabilities well beyond basic phylogenetic analysis. Some of those new capabilities are discussed in Chapters 12–15 of this book.

Although MEGA 5 runs smoothly on Macintosh and Linux platforms, its "look and feel" is very Windows-like. Chapter 11 goes into more detail about some platform-specific issues. Appendix II, "Additional Programs," briefly discusses both PAUP* and PHYML and where to obtain them.

Despite continued software development, no single program is most suitable for all of the methods discussed. A forthcoming major upgrade of MrBayes to version 3.2 will add some very desirable features to this program. Readers who are already familiar with MrBayes 3.1 are strongly encouraged to read Chapter 10 in order to take advantage of these new features, which will make their lives easier.

Just What Is a Phylogenetic Tree?

To a mathematician, a phylogenetic tree is an abstract construct embodied in a special type of directed or undirected graph. To systematists and most evolutionary biologists, a tree is a representation of the relationships of different species to their ancestors. To a molecular biologist, a tree is a representation of the relationships of gene or protein sequences to their ancestral sequences. All agree, however, that a tree consists of nodes connected by branches (except that mathematicians tend to refer to branches as edges).

The tips of a tree, sometimes referred to as **leaves**, are the **external nodes** and biologically they represent existing taxa. Systematists and evolutionary biologists generally think of taxa as being synonymous with species, while molecular biologists think of taxa as being synonymous with sequences. In either case, these leaves/taxa/sequences are the only entities in a tree that we can be sure about. They represent the real data—factual information—from which everything else in the tree is inferred. That factual information can be the states of morphological characters or it can be the states of nucleotide or amino acid characters in macromolecular sequences. The discussion in this book will be limited to molecular sequence characters.

Even with sequence data, however, much is assumed. We may use sequences of the cytochrome c gene to make a tree in which each sequence is from a different species, so we may refer to the tips as the "cow sequence," "dog sequence," etc. In fact, the cow sequence is not the cytochrome c sequence for all cows; it is the sequence of that gene from one particular cow. We can use that sequence to represent all cows because we assume (1) that any variation in cow sequences occurred after cows diverged from dogs, and (2) that any one randomly selected cow sequence is as representative of cows as the next. Both these assumptions arise from a fundamental assumption of phylogenetics, that of genetic isolation—that there is no genetic exchange between taxa. When this fundamental assumption is violated, different parts of the information used to make the tree (gene sequence, set of sequences, full genome) can have different histories and valid trees cannot be constructed.

The interior nodes of the tree represent *hypothetical* ancestors. We don't know the character states of those ancestors and in most cases cannot know those characteristics because the ancestors no longer exist. We can only infer the characteristics of ancestral sequences from information about their descendants (the leaves of the tree). We assume a process of modification with descent in which mutations accumulate and are inherited by their descendants. **Branches** connect nodes and represent that descent, and **branch lengths** represent the amount of change between an ancestor and its descendant.

The most important fact to keep in mind about any phylogenetic tree is that it is not fact, it is an estimate—an inference. Many evolutionary biologists would say that a tree is a hypothesis, but I disagree. A hypothesis is useful only if it can be tested, and we cannot test the validity of evolutionary trees. We can estimate trees, we can estimate how accurate they might be, and we can change our estimates (and the trees) when new data are available, but that's it. An estimate is a fine and useful thing; just keep in mind that it is not absolute truth.

Estimating Phylogenetic Trees: The Basics

Chapters 2–11 of this book cover the basics of four distinct and equally important steps that are involved in making a phylogenetic tree based on molecular sequence data:

1. Identify and acquire the sequences that are to be included on the tree.

2. Align the sequences.

3. Estimate the tree by one of several methods.

4. Draw the tree and present it to an intended audience.

Chapters 2–9 describe implementing these four steps using the MEGA 5 software package. The data acquisition, alignment, and tree-drawing functions of MEGA 5 are so elegantly implemented and easy to use that this is the program of choice for most phylogenetic methods. Understanding Chapters 2–9 will permit confident estimation of valid phylogenetics trees by a well-accepted, reliable method.

Chapter 2 is a tutorial that takes the reader through each step, using MEGA 5 to construct a simple phylogenetic tree. The primary purpose of Chapter 2 is to familiarize you with both the mechanics of implementing the steps necessary to make a tree and the basics of the MEGA 5 software. Chapter 2 will *not*

TABLE 1.1 Some Conventions Used in this Book

CONVENTION	DESCRIPTION
Click	Use the mouse to position the cursor over the indicated button on the screen (as in "click the OK button"), then depress and quickly release the mouse button.
Double-click	Click twice rapidly without moving the mouse.
Drag	Position the mouse and, while holding down the mouse button, move the mouse to another position.
Select	Highlight a menu item, section of text, or an object on the screen by dragging the mouse over or clicking (or double clicking) on the desired operation. In the text these operations are indicated in **Bold Sans Serif** type.
Enter	For command line programs such as MrBayes and when using this book's utility programs, text shown in the `Courier typeface` indicates commands that you will type into an input file (or see on the screen if you have downloaded a utility file).

lead to the optimal tree based on the data provided, so details of each step are covered in the subsequent chapters.

Chapter 3 deals in more detail with Step 1: identifying the sequences that might be included on a tree, deciding which to include and which to exclude, and downloading those sequences from international databases operated by government agencies..

Chapter 4 deals with the critical problem of aligning sequences (Step 2). Sequence alignments, whether of nucleotides or proteins, are the data upon which phylogenetics programs operate to estimate a tree. If you don't get this part right, nothing you do in later steps will matter; your tree will be worthless.

Chapter 5 very briefly discusses and compares the major methods for estimating phylogenetic trees. Chapter 6 then deals with Neighbor Joining (NJ), the most widely used tree estimation method, and discusses the often intimidating issue of bootstrapping to assess tree reliability. In many ways, Chapter 6 is the meat of this book. The major phylogenetic concepts are covered in this chapter. Subsequent chapters dealing with other phylogenetic methods are mainly technical and simply illustrate how to accomplish the same goals with other methods and software.

Chapter 7 describes how to draw trees in various ways and how to choose which style will make it easiest for your audience to correctly interpret the tree and understand the biological point you are making.

Chapter 8 discusses using MEGA to estimate trees by Maximum Parsimony (MP). MP is one of the first methods to have been applied to molecular sequence data.

Chapter 9 describes how to use MEGA to estimate trees by Maximum Likelihood (ML). ML is a statistical method that has long been appreciated, but is not widely used outside of the fields of systematics and phylogenetics because it is perceived as being intimidating, slow, and not applicable to large data sets. Recent advances in software have solved the speed problem and have made ML applicable to large data sets. I hope that after reading Chapter 9 you will find ML no more intimidating than NJ or MP.

Chapter 10 discusses constructing trees by Bayesian Inference (BI) using MrBayes. Like ML, BI is a powerful statistical method and it has proven to be slightly more accurate than the other methods (Hall 2005).

Chapter 11 discusses the vagaries of various platforms and operating systems so that whatever platform is available, you will be able to use it efficiently for phylogenetic analysis.

Beyond the Basics

Those readers without previous experience with phylogenetics will probably find the information in Chapters 2–11 sufficient to meet their needs for some time to come. Increased experience with phylogenetics, however, often leads to recognition of the utility of more advanced aspects of phylogenetic analysis. Some of those advanced topics are discussed in Chapters 12–15.

Chapter 12 returns to the critical topic of sequence alignment and introduces a new tool for improving those alignments, the Web-based program GUIDANCE. GUIDANCE allows you to assess the quality of an alignment and to identify those regions that degrade the alignment's quality. Even more importantly, the program permits you to identify sequences that should *not* be part of the alignment—i.e., sequences that may not be homologs of the other sequences to be included in a tree. Rooting a phylogenetic tree requires the use of one or more "outgroup" sequences that are more distantly related to an "ingroup" of sequences than the ingroup sequences are to each other (see Chapter 7, p. 108). Outgroup sequences thus must be related to the taxa of interest only distantly—but not so distantly that the sequences are nonhomologous. GUIDANCE is very helpful when deciding whether or not to designate a particular sequence as an outgroup sequence.

Chapter 13 deals with the reconstruction of the amino acid sequences of ancient ancestral proteins and nucleotide sequences of ancestral genes. This subject, which some call *paleobiology*, permits the estimation of ancestral sequences as a necessary prelude to actual synthesis of ancestral proteins. Biochemists and others interested in understanding protein function often have an interest in synthesizing, then studying, a protein they believe to be the common ancestor of a group of proteins with distinct, but mechanistically related, functions. The first step in the process is to estimate a reliable phylogenetic tree, and the second step is to estimate the most likely sequence of that common ancestor.

Chapter 14 deals with another advanced topic, detecting *adaptive evolution*—i.e., purifying or diversifying selection—along phylogenetic trees.

Chapter 15 concerns phylogenetic *networks* as opposed to phylogenetic trees. In some circumstances phylogenetic trees are an inadequate means of describing the historical relationships of taxa. For instance, trees based on different genes from the same set of species are often significantly different because those genes have different evolutionary histories as the result of recombination, horizontal transfer, or other "reticulate" events. In such cases phylogenetic networks more realistically describe evolutionary history. An interior node of a phylogenetic tree may have multiple nodes descending from it, but it can have only one immediate ancestral node. In a phylogenetic network, a node can have both multiple descendants and multiple immediate ancestors.

Learn More about the Principles

Just as it is possible to implement molecular methods without understanding them by following the protocols in commercial "kits," it is also possible to implement phylogenetic methods without understanding them by following the protocols in this book. Most of us insist that our students understand the principles underlying the methods implemented by these kits, however, because we know that without such an understanding it is impossible to spot and troubleshoot many problems.

It is in this spirit that the reader will find *Learn More* boxes scattered throughout the text. These boxes present somewhat more detailed background on the biological and/or mathematical principles underlying the various methods and suggest further reading. It is not necessary to read the boxes to be able to estimate a valid phylogenetic tree, but understanding these principles can help troubleshoot the phylogenetic problems that arise when estimating trees from molecular alignment data.

Readers who want to go beyond the Learn More boxes will find Dan Graur and Wen-Hsiung Li's *Fundamentals of Molecular Evolution* (2000) and Masatoshi Nei and Sudhir Kumar's *Molecular Evolution and Phylogenetics* (2000) very helpful and enjoyable. Li's *Molecular Evolution* (1997) and Chapters 11 (Swofford et al. 1996) and 12 (Hillis et al. 1996) of *Molecular Systematics* (David Hillis, Craig Moritz, and Barbara Mable, eds.) provide more detailed insights into these topics. Joe Felsenstein's outstanding book, *Inferring Phylogenies* (2004), is very technical, very mathematical, and not for the faint of heart. At the same time, it is delightfully written and well worthwhile for anyone who really wants to understand phylogenetic theory. *Phylogenetic Networks* by Daniel Huson, Regula Rupp, and Celine Scornavacca (2011) provides an extremely detailed explanation of the concepts, algorithms, and applications of this relatively new approach (described briefly in Chapter 15 of this book). Like Felsenstein's book, *Phylogenetic Networks* is highly technical and mathematical, but it is also a very worthwhile presentation of this recent alternative to phylogenetic trees for understanding relationships among taxa.

About Appendix III: F.A.Q.

Well, not really "Frequently Asked Questions," but "Sometimes Asked Questions." Or more likely, "Questions That Should Be Asked but Probably Aren't."

This book is organized and presented as though you will use MEGA whenever possible for every step in the process of estimating and presenting a tree. What if you want to use MEGA to download sequences, but you want to use another program for alignment? What if you have estimated a tree using other programs, but you want to use MEGA to draw and present that tree? The more you use phylogenetics the more likely you are to want to explore and use alternative programs. Such issues can interrupt the flow of an individual chapter and be distracting, so I have collected them in Appendix III. Look there first for tips and hints. I hope that these "What if?" questions will both help and encourage you to expand your horizons beyond those presented in this book.

Computer Programs and Where to Obtain Them

Programs are continuously being updated. Readers should always download the latest versions from the sources listed below. Readers should also be aware that new versions might have screens that differ from the examples in this book, or they may accept commands in slightly different ways. Usually the modifications are minor and will present few problems to the careful user.

MEGA 5

MEGA 5 (Tamura et al. 2011) is a modern, integrated phylogenetics package that elegantly handles downloading sequences through its own specialized web browser; aligning sequences through its implementations of ClustalW and MUSCLE; estimating phylogenetic trees by a variety of methods including Neighbor Joining (NJ), Maximum Parsimony (MP), and Maximum Likelihood (ML); and drawing those trees in a variety of ways. MEGA 5 is free and can be downloaded from http://www.megasoftware.net.

MrBayes

MrBayes is used for Bayesian Inference (BI) of phylogenetic trees (Ronquist and Huelsenbeck 2003). The MrBayes home page (http://www.mrbayes.net) provides helpful information and links to the download site. MrBayes is free and is available for Macintosh and Windows computers. The source code is available for compilation on Unix- and Linux-based systems. On Macintosh OSX computers, it is often more convenient to compile the source code and then to use MrBayes through the Terminal program (see Chapter 11).

Release of MrBayes 3.2 was imminent at the time of this book's publication, and it is this release that is described in Chapter 10.

FigTree

FigTree is an excellent tree-drawing program. It reads tree files in both Nexus and Newick formats. It also has an extended Nexus format that includes such features as fonts, branch colors, etc. FigTree is the default format for drawing the consensus tree in MrBayes 3.2. Like MEGA's tree-drawing program, Fig Tree permits re-rooting, rotating clades around a branch, showing or hiding branch lengths, and other features. Download from http://tree.bio.ed.ac.uk/software/figtree/.

Codeml

Codeml is part of the free PAML (Phylogenetic Analysis by Maximum Likelihood) package. You can download PAML for Macintosh, Windows, and Unix/Linux from http://abacus.gene.ucl.ac.uk/software/paml.html. The current version of PAML is 4.4b.

SplitsTree and Dendroscope

SplitsTree and Dendroscope are free programs for estimating and drawing phylogenetic networks. They are available from http://www-ab.informatik.uni-tuebingen.de/software/splitstree4/welcome.html and http://www-ab.informatik.uni-tuebingen.de/software/dendroscope/welcome.html, respectively.

Utility Programs

I have provided a couple of utility programs that are very helpful for changing file formats and for reconstructing the sequences of ancestral proteins and nucleic acids. Those programs, along with various examples, templates, and so forth can be downloaded from the *Phylogenetic Trees Made Easy* website (http://sites.sinauer.com/hall4e/). Chapter 11 gives details on installing and using these programs.

Text Editors

Users who intend to use MrBayes, codeml, or the utility programs will need a text editor program to prepare the necessary input files. Word processor files such as those written by Microsoft Word, WordPerfect, etc. will not work well.

- For Macintosh I recommend the free TextWrangler from BareBones Software (http://www.barebones.com/) or the more sophisticated commercial BBEdit from the same source.
- jEdit is another excellent free text editor, and the Java-based version will run on any platform. Go to http://www.jedit.org/index.php?page=download and choose the Java-based installer.
- KomodoEdit is a free text editor from ActiveState and is available for Windows, Linux, and Macintosh (http://www.activestate.com/komodo-edit/downloads).

Acknowledging Computer Programs

Phylogenetic analysis is completely dependent upon computer programs; without those programs, molecular phylogenetics simply would not be possible. The programs described in this book are free—at least they are free to us, the users. The authors of those programs have paid dearly for them in terms of time, effort, and creativity. The only reward the authors get for their effort (and it is an enormous effort) is having that software cited in publications. Please be sure to cite the software you use just as rigorously as you cite published papers. You can find the appropriate citation in the program itself, in the documentation, or at the program's website. If you can't find a paper, simply cite the program's website. If you find something about a program particularly useful, or just a pleasure to use, it would be nice to drop the authors an email. Authors who feel appreciated tend to keep improving their software, making all of our lives easier.

The *Phylogenetic Trees Made Easy* Website

The *Phylogenetic Trees Made Easy* website was created especially for this book. Located at http://sites.sinauer.com/hall4e/, the site contains a set of files that will allow you to follow Chapters 2–10 without actually downloading the sequences. It also includes copies of many of the relevant output files so you can compare your results with those I obtained in the event that you have difficulties. The files have been compressed into Zip archives to speed downloading and circumvent some institutional firewalls; these files will need to be extracted with appropriate software. Most computers will have the necessary software already installed, but if yours does not, download one of these free programs: Stuffit Expander for Mac and for Windows (http://www.stuffit-expander.com/ stuffit-expander.html), or FreeZip for Windows (http://members.ozemail. com.au/~nulifetv/freezip/). Linux users should use the `tar` command to extract the files.

The downloaded folder also includes templates, or boilerplate, for various input files. You can copy those templates directly into your own input files in lieu of typing everything manually. It is amazingly easy to make a tiny typing error, such as `1st _pos` instead of `1st_pos`, and then spend hours trying to figure out why your program won't run. Copying blocks of critical text, then modifying those blocks for your specific needs, is a good way to avoid such problems. Indeed, many of us use such templates routinely. Throughout the text these files and templates are indicated with an icon:

 Chapter 2 files: Tutorial.mas

This icon means in the folder that you downloaded, in the subfolder Chapter 2 files, use the file named Tutorial.mas.

You can also register for updated information at this website. If you choose to do this you will receive email announcements of updates to programs, updated templates for MrBayes files, and so forth. No advertisements will be sent to you, nor will your information be passed on to others.

Tutorial: Estimate a Tree

Why Create Phylogenetic Trees?

Today phylogenetic trees appear frequently in molecular papers that are unrelated to phylogenetics, or even to evolution *per se*. Their inclusion reflects the growing recognition of trees as a tool for understanding biological processes. Phylogenetic trees allow you to organize your thinking about a protein of interest in terms of its relationship to other proteins, and may allow you to draw conclusions about a protein's biological functions that would not otherwise be apparent.

Because genomes are being sequenced at increasingly rapid rates, our knowledge of the DNA and amino acid sequences of proteins is far outstripping our knowledge of their biological and biochemical functions. As a result, we are frequently forced to assign functions to proteins on the basis of sequence homology alone. As the sequence databases grow larger, more and more functions are assigned based on a sequence's homology to other sequences whose functions have been only tentatively assigned based on their homology to still other sequences. Examination of a phylogeny can allow you to determine how closely your sequence relates to a sequence whose function is *actually known* from biological or biochemical information.

As covered in this book, there are five steps involved in estimating a phylogenetic tree based on molecular sequence data:

1. Identify sequences that are homologous to your sequence of interest.

2. Decide which of those homologous sequences to include on the tree.

3. Download electronic files of those sequences.

4. Align the sequences.

5. Use the resulting alignment to estimate a phylogenetic tree.

The first three steps require a computer that is connected to the internet and is running a set of suitable programs. In addition to guiding you through these steps, this manual will suggest suitable programs and provide information on obtaining them.

About this Tutorial

The software that forms the core of this book is MEGA 5 (Molecular Evolutionary Genetics Analysis, Version 5), and the purpose of this tutorial is to familiarize you with the core features of MEGA 5 while illustrating the basic steps of creating a phylogenetic tree. The strategy for estimating the tree illustrated in this example is far from optimal, and subsequent chapters will illustrate details that can vastly improve on the approach in this chapter. Nevertheless, the example that follows here will help you learn how the basic features of MEGA 5 work, and the tree that results—as far as it goes—will be valid.

Before you begin this tutorial, you will need to download and install MEGA 5 (see p. 8). If you don't have MEGA 5 yet, or if the preceding sentence makes no sense to you, please read Chapter 11 now.

If you have the earlier version, MEGA 4, you should replace it with MEGA 5. Mega 5 has several new capabilities that are not available in MEGA 4 and the interface is significantly different. MEGA 5 includes an "Update" button in the main window, and you should use that button frequently to ensure that you are using the latest version. Some updates may just be bug fixes, but others will add new capabilities or extend old ones. When new capabilities are introduced those of you who register on the PTME4 website (see pp. 9–10) will be notified by email and I will post an appropriate description of the new capability to the PTME4 website.

MEGA 5 includes an excellent context-sensitive **Help** menu that provides detailed background information about each of the functions it performs. There is also an excellent manual that includes tutorials on the major MEGA 5 functions. I strongly advise you to take advantage of the MEGA 5 manual and tutorials. (From this point forward I will simply write MEGA, rather than MEGA 5, but I am always referring to MEGA 5.)

Macintosh and Linux users

MEGA has the look of a Windows application, which is all right because that allows an identical interface on all three platforms. Still, there are some platform-specific issues that you should be aware of. To avoid unnecessary frustration, before starting this tutorial Macintosh and Linux users should read about the platform-specific issues covered in Chapter 11.

A word about screen shots

This tutorial will have the most value if you and your computer follow along at every step. In several places the descriptions assume that you are doing so, and therefore do not include screen shots of every detail. What you do see on your computer screen as you follow along may differ from the screen shots shown in this book. The software being discussed may have changed since this book was written, or websites may have changed the way they display information. Websites such as BLAST that show information from databases are particularly likely to differ from the illustrations. New sequences are constantly being added to the databases, making it highly likely that you won't see exactly what is in the illustration. Don't worry about such differences. Details may differ, but the point being illustrated will remain essentially the same.

If the interface of a software package has changed from what is shown in the book, just look about the program a bit. You probably will be able to find the same function elsewhere. If necessary, consult the documentation for the software.

Search for Sequences Related to Your Sequence

Quite likely you have a particular protein or nucleic acid sequence you are interested in and you need to find other sequences that are related to it. By "related," we mean another sequence that is sufficiently similar to the sequence of interest that it is likely the two sequences share a common ancestor (i.e., the sequences are related by descent).

The easiest (and the safest) way to find related sequences is to enter your sequence of interest and search the international nucleic acid and protein databases for similar sequences. You can do the entire search on the World Wide Web courtesy of government computers. The search-and-download program used in this tutorial is the National Center for Biotechnology Information's BLAST (Basic Local Alignment Search Tool). BLAST uses your sequence as a "query" to search the world's combined databases of protein and nucleic acid sequences. I will assume that your sequence exists as some sort of electronic file—perhaps as a simple text file or as a file from a sequence manipulation program. Almost any format will do. The example I use here is that of an α-glucuronidase gene from the bacterium *Thermotoga petrophilia*, strain RKU1.

Start MEGA to see the main window. From the **Align** menu, choose **Do BLAST Search** (**Figure 2.1**).

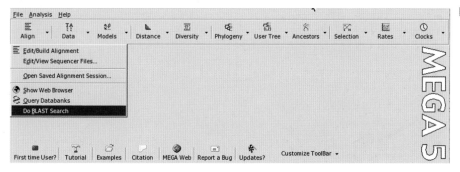

Figure 2.1

MEGA's built-in web browser automatically takes you to the NCBI BLAST page (**Figure 2.2**). That page is divided into four sections, each of which requires an action on your part. In the top section there is a text box labeled **Enter accession number, gi, or FASTA sequence** into which you enter a **query sequence**, which is the sequence for which you want to find homologs to include on your tree. You can either copy and paste this sequence or enter an accession number or a gi number for a sequence that is in Genbank. In this example we will use the accession number for the *Thermotoga petrophilia* complete genome sequence. Enter CP000702 in the text box. Since this number specifies the entire genome sequence, we must also specify the region of the genome that encodes our gene of interest, that for α-glucuronidase. The two boxes to the right of the text box are labeled **Query Subrange**. In the upper box enter 1005082, and in the lower box enter 1006524; these numbers specify the range of nucleotide positions encoding the α-glucuronidase gene.

Figure 2.2

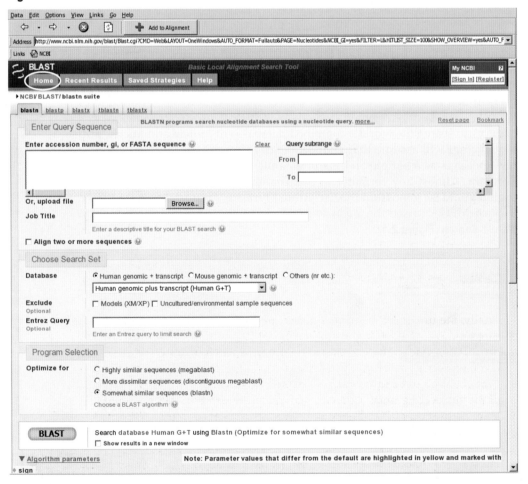

The next section down is the **Choose Search Set** section (see Figure 2.2). The default is to search the **Human Genomic + transcript (Human G+T)** database, which certainly will not include genes homologous to our bacterial query sequence. From the pull-down menu in that section choose **Nucleotide collection (nr/nt)**.

The next section is the **Program Selection** section, where you choose which BLAST program you will use for the search. This is not a trivial matter, so don't just accept the default (which is **megablast**). Use **megablast** when you only want to find very closely related sequences; use **discontiguous megablast** to cast your net a little wider; and use **blastn** to find even distantly related sequences. For this tutorial, choose **blastn**, but don't hesitate to try the other options to explore the extent to which they constrain your search—exploring is good! The page should then look like **Figure 2.3**.

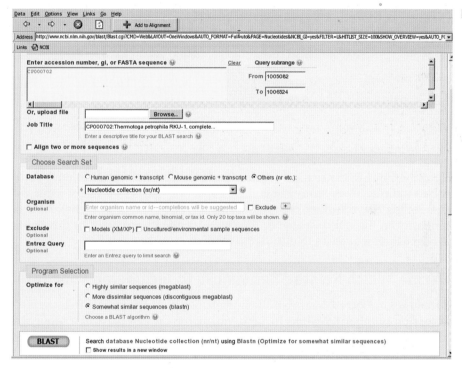

Figure 2.3

Finally, the unlabeled bottom section contains the **BLAST** button that starts the search. Click that button and BLAST will tell you that it is searching the database; eventually it will display the **Results** page (**Figure 2.4**).

The Results page displays a graph that summarizes the results of the search—the "hits," or **subject sequences**. The bars are color-coded to reflect the similarity of the subject alignment sequences to the query sequence, with red indicating alignment scores >200. The query sequence appears at the top (the thicker red bar directly below the key), with the subject sequences aligned below. The length and position of a subject bar indicates the region of similarity to the query. The long red bars indicate subject sequences that align with high similarity

Figure 2.4

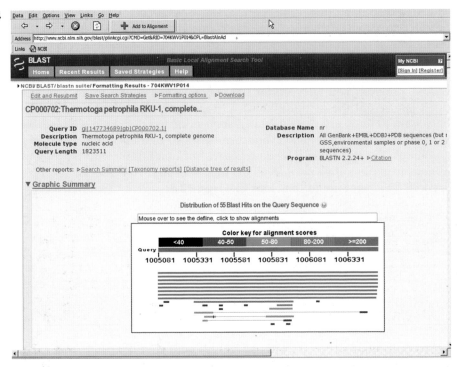

along the entire query sequence. Below these long bars, different-colored shorter bars indicate lower-scoring alignments over only part of the query sequence.

Scrolling down further, we see the subject sequences listed in the order of similarity to the query sequence. We will use the information in this list to decide which sequences to include in our tree.

Decide Which Related Sequences to Include on Your Tree

Each entry in **Figure 2.5** consists of eight elements. At the far left, in blue, is an **accession number** that links to the Genbank file for that particular sequence. Next is a brief description of the sequence.

The next two columns show the Maximum (Max) score and the Total score, respectively. The higher these scores, the more closely related that subject sequence is to the query sequence. When the Max score is different from the Total score, it means that more than one portion of the subject sequence aligns meaningfully with the query sequence; the **Max score** is the score for the highest-scoring segment of the subject sequence, and the **Total score** is the sum of the scores for all the segments that align (including noncontiguous alignments). The Max score for the first entry, for example, is 2775 and the Total score 3151.

The column labeled "Query Coverage" shows how much of the query sequence aligns with the subject sequence—essentially the same information provided by the lengths of the bars in Figure 2.4.

Figure 2.5

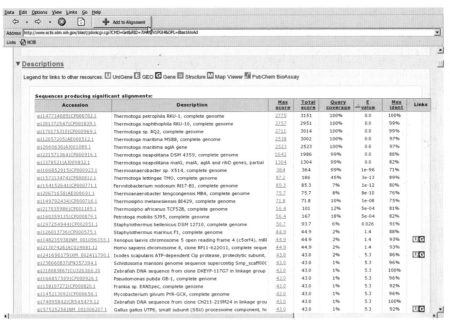

The next column shows the **expectation value**, or **E value**, a parameter that describes the number of hits with scores this high that one would "expect" to see by chance when searching a database of a particular size. For example, a hit with an E value of 1 can be interpreted as meaning that, in a database of the size being searched, we might expect to see 1 match with a similar score simply by chance. The closer the E value is to zero, the more significant the score and the more likely it is that the hit is a homolog of the query sequence. In the lower half of Figure 2.4 you will see E values >1.0, so the E value is obviously not exactly the same as the probability of the match being due to chance rather than to homology, but it is useful in judging whether the hit is likely to be a homolog of the query. (I discuss scores and their E values in more detail in Chapter 3.)

The **Max Ident** column shows the identity between the query sequence and the portion of the subject sequence that is most identical to the query.

We will use all of the information in these columns to choose which of the hit sequences to include on the tree.

Establishing homology

It is important to understand that there are no hard-and-fast "right" choices in deciding which sequences to include; those choices depend upon the purpose for which you are estimating the tree. However, the probability of a hit sequence being a homolog of the query sequence is an important issue because the tree should include *only* homologous sequences.

It is a basic assumption of phylogenetics that all sequences (or organisms) on a given tree are descended from a common ancestor. Indeed, in evolutionary studies the word *homologous* means "descended from a common ancestor."

This issue will be discussed in more detail in Chapter 3, but for the moment we will include only those sequences we can be confident are homologs of the query sequence; thus we will set the E value cutoff parameter at values <1e-3 (i.e., we will include on the tree only those sequences whose E value is <10^{-3}). Likewise, we probably don't want to include sequences that are homologous over just a small portion of the query. For this exercise we will only include sequences whose query coverage is ≥60%.

   Now that we have decided which sequences we *can* include, we need to decide which sequences we actually *want* to include.

To include or not to include, that is the question

To decide whether or not we want to include a given subject sequence, we need to look at the alignment between that sequence and the query. To do that we can click on the **Max Score** for a sequence. **Max Score** is a link that takes us down to the alignment. Clicking on the score for *Thermotoga petrophilia* RKU-1 takes us to its alignment (**Figure 2.6**).

   Each alignment description begins at the top left with the accession number that links to the database file for the subject sequence and, to the right of the accession number, a description of the file.

Figure 2.6

Below that there is a line that reads "Features in this part of the subject sequence," followed by a link that reads "alpha-glucuronidase. Glycosyl Hydrolase Family 4." That link will be our pathway to the sequence we want to add to our tree.

Next there is a section that describes the properties of the alignment. We see that its score is 2775, its E value is 0.0, and its Identity is 1443/1443, meaning that the query and subject sequences are identical at all 1443 positions without any gaps (i.e., it is a perfect match). This is not surprising, since the query sequence came from exactly that strain. We also note the message "Strand = Plus/Plus." That means that the sequence in the Genbank file is the same as the query sequence. "Strand = Plus/Minus" would mean that the Genbank file sequence is the reverse complement of the query sequence.

Looking at the alignment itself, we see that each line is identified as being Query or Subject sequence and that the ends are labeled with the base numbers in the respective sequences. Vertical lines connect bases that are identical in the subject and query sequences, so a quick scan of the sequence gives an impression of the degree to which the query and subject sequences match.

Scrolling down, immediately after this alignment we see a second alignment with a different "features" link (**Figure 2.7**). For that alignment, the identities are 195/259; i.e., only 259 bases of the 1443-base query sequence align with this portion of the *T. petrophila* genome. Although the E value of this second portion of the genome (shown as "Expect = 5e-26") meets our 10^{-3} E value

Figure 2.7

cutoff parameter, we will not include it because it aligns with only about 18% of the query—far below the ≥60% parameter we established for inclusion.

If we continue scrolling down, we see four additional alignments to this genome. It seems that *T. petrophila* strain RKU-1 has, in addition to the query sequence itself, four other regions with homology to the query. None of these align over the requisite 60% of the query length, and the last one has an E value of 1.4, so we will only include the first sequence (i.e., that in Figure 2.6) on the tree.

Returning to the list of **Sequences producing significant alignments** (see Figure 2.5), after analyzing each of these the sequences more closely as described for Figures 2.6 and 2.7, we find that only the first eight* sequences in Figure 2.5 meet our E value and query coverage criteria for inclusion on the tree.

Download the Sequences

Having chosen the sequences we want to include on our tree, next we need to download each of these eight sequences to MEGA's Alignment Explorer. For each sequence in turn:

---

*Note that your screen may show a different number of sequences that meet the criteria; this is because sequences have been added to the database since this book was written.

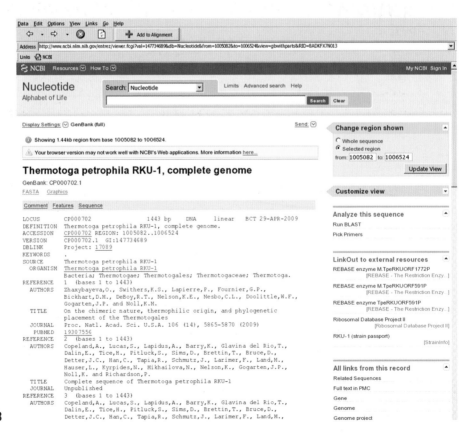

Figure 2.8

1. Click the **Max Score** link to go down to the alignment. Be sure to note whether the alignment is **Strand = Plus/Plus** or **Strand = Plus/Minus**.

2. Click the "features" link (circled in Figure 2.7). That link takes you to the corresponding Genbank file (**Figure 2.8**). Notice that although the Genbank file indicates it is that of the complete genome, the beige **Change Region Shown** section shows that only the **Selected region** (i.e., from 1005082 through 1006524) is shown.

3. If the query strand and the subject strand are in different orientations (Plus/Minus or Minus/Plus), click the black triangle next to **Customize view**, tick the **Show reverse complement** box, then click the **Update View** button (**Figure 2.9**). Notice that in Figure 2.9, which is the file for the second sequence in the list (see Figure 2.5), the first number in the sequence range is larger than the second number—a clear indication that the sequence is that of the minus strand. It is a common error to add sequences without noticing which strand is being added. Don't do that!

Figure 2.9

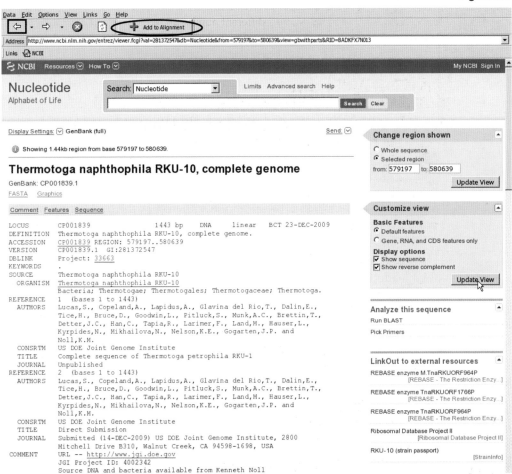

4. If you had to **Show Reverse Complement**, click the **Update View** button so that the correct sequence is displayed at the bottom of the page. Scroll down to the bottom of the page and look at the sequence to be sure that it is indeed the sequence you want.

5. Click the **Add to Alignment** button displaying a red cross that is near the top of the browser window (circled in Figure 2.9). Doing that displays MEGA's **Alignment Editor** window with the sequence added to it. Overlapping that window will be a small **Information** window that informs you that the sequence has been added (**Figure 2.10**). Click the **OK** button to dismiss the **Information** window.

Figure 2.10

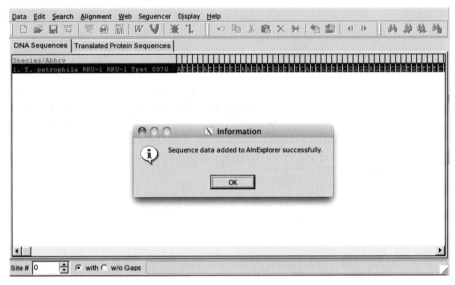

6. Click on MEGA's **Web Browser** window, then click the back arrow near the top left of the window (boxed in Figure 2.9) to get back to the page that lists the sequences you will add to the tree.

Repeat steps 1–6 for each of the eight chosen sequences. When you have downloaded all eight, the Alignment Explorer window should look something like **Figure 2.11**.

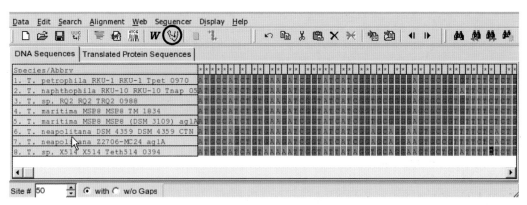

Figure 2.11

This is a good time to save the contents of the Alignment Explorer in order to avoid losing all your work. From the **Data** menu of the **Alignment Explorer** window, choose **Save Session** (alternatively, you can type `Control-s`). Navigate to an appropriate folder and choose a file name with the extension `.mas` to indicate an alignment file. For the purpose of this tutorial, just name it `Tutorial.mas`. We are now done with the web browser, so you can close that window.

 Chapter 2: Tutorial.mas

Align the Sequences

As mentioned earlier, it is a basic premise that all sequences on a tree are homologous. All tree-building methods also assume that, in the set of homologous sequences, all of the bases in a column are also homologous (i.e., the current bases are all descended from the base that was present at that position in the ancestral sequence). If no insertion or deletion mutations—collectively called "indels"—ever occurred, simply writing the sequences one above the other would accomplish that end. However, indels *do* occur, and they change sequence lengths, shift the positions of bases, and affect the sequence of amino acids. **Alignment** is a process designed to introduce gaps into the sequence so as to shift bases back to their corresponding homologous positions. This process, covered in detail in Chapter 4, is so important that the quality of a phylogenetic tree can be no better than the quality of the sequence alignments used to produce the tree.

In the **Alignment Explorer** window click the **Align Selected by Muscle** icon (circled in Figure 2.11), and choose the **Align DNA** option (alternatively, you can choose **Align by Muscle** from the **Alignment** menu). A dialog will remind you that nothing is selected and ask if you want to select all sequences. Click the **OK** button. A new **Muscle Parameters** window is displayed. This window will be discussed in detail in Chapter 4, but for this tutorial you can just click the **Compute** button.

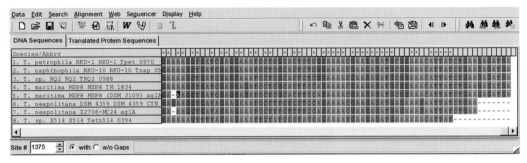

Figure 2.12

   A progress window will briefly appear, after which the Alignment Explorer window will display the aligned sequences. Use the horizontal scrolling bar at the bottom of the window to scroll to the right, to the end of the alignment so that it looks like **Figure 2.12**.

   Click within the alignment on a base that is to the right of some gap and notice that the number of that site is shown at the very bottom of the Alignment Explorer window. That number, by default, shows the numbering with gaps (i.e., the number from the beginning of the alignment). To see the base as numbered from the beginning of the *sequence*, tick **w/o Gaps**.

   The alignment is now complete, and it is important to save the alignment again. Save it as `TutorialAligned.mas`.

 Chapter 2: TutorialAligned.mas

Make a Neighbor Joining Tree

The MEGA function that estimates phylogenetic trees from the alignment cannot use the `TutorialAligned.mas` file directly; it requires a file in the MEGA format. From the **Data** menu of the Alignment Editor choose **Export**, and in the submenu choose **MEGA Format**. Name the file `TutorialAligned.meg` to indicate a MEGA file. A dialog box will ask for the title of the data. It doesn't matter what title you enter; `Tutorial DNA sequences` will be just fine. A second dialog will ask you if these are protein-coding sequences. It is important to click the **Yes** button.

 Chapter 2: TutorialAligned.meg

Figure 2.13

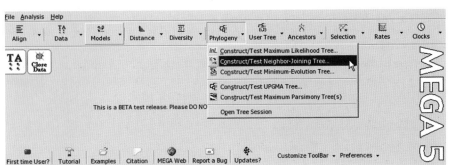

Figure 2.14

We are now done with the **Alignment Editor**, so you can close that window. You will be asked if you want to save the alignment, but you have already done that, so click **No**.

In MEGA's main window, choose **Open a File/Session** from the **File** menu (**Figure 2.13**), navigate to the folder where you saved `TutorialAligned.meg`, then select and open that file.

The MEGA main window now displays two new large buttons in the main portion of the window (**Figure 2.14**). From the **Phylogeny** menu choose **Construct/Test Neighbor Joining Tree**. A dialog box appears to ask **Would you like to use the currently active data file (TutorialAligned.meg)**. Click the **Yes** button.

The **Analysis Preferences** window opens (**Figure 2.15**). This window lets you set the conditions for all analyses performed by MEGA. In this case, it tells us that we want to do **Phylogeny Reconstruction** by the **Neighbor Joining** method.

Options Summary	
Option	Selection
Analysis	Phylogeny Reconstruction
Scope	All Selected Taxa
Statistical Method	Neighbor-joining
Phylogeny Test	
Test of Phylogeny	None
No. of Bootstrap Replications	*Not Applicable*
Substitution Model	
Substitutions Type	Nucleotide
Genetic Code Table	*Not Applicable*
Model/Method	Maximum Composite Likelihood
Fixed Transition/Transversion Ratio	*Not Applicable*
Substitutions to Include	d: Transitions + Transversions
Rates and Patterns	
Rates among Sites	Uniform rates
Gamma Parameter	*Not Applicable*
Pattern among Lineages	Same (Homogeneous)
Data Subset to Use	
Gaps/Missing Data Treatment	Complete deletion
Site Coverage Cutoff (%)	*Not Applicable*
Select Codon Positions	☑ 1st ☑ 2nd ☑ 3rd ☑ Noncoding Sites

| ✓ Compute | ✗ Cancel | ? Help |

Figure 2.15

Choices that can be modified are shown in yellow. I will deal with these choices extensively in Chapter 6, but for purposes of this tutorial simply click the **Compute** button.

A **Tree Explorer** window opens to display the Neighbor Joining tree (**Figure 2.16A**). Choose **Save Current Session** from the **File** menu to save the tree (**Figure 2.16B**). Save the tree using the file name `Tutorial.mts`.

Figure 2.16A

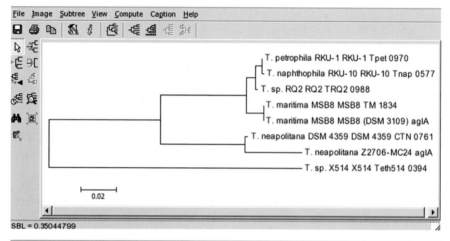

Figure 2.16B

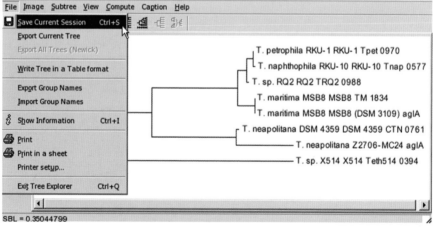

If you want to print the tree and you are working on a Windows computer, choose **Print** from the **File** menu. If you are working on a Macintosh or Linux computer, you cannot print directly from MEGA. Instead choose **Save as PDF file** from the **Image** menu (**Figure 2.17**). This will save the tree as a PDF file that can be printed by Preview (Macintosh) or Adobe Acrobat Reader (Linux or Macintosh).

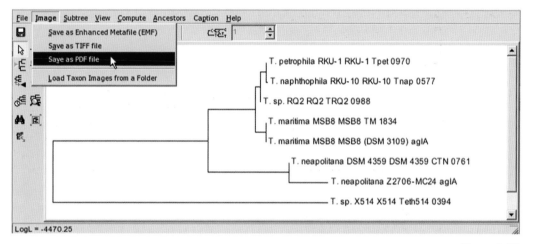

Figure 2.17

That's it! You have made a valid, if small, phylogenetic tree. Chapter 7 will discuss a variety of ways you can modify the appearance of the tree so that it presents the information to your audience in exactly the way that is most helpful to them. Notice that the bulk of time and text was devoted to choosing which sequences to include, and then downloading those sequences from Genbank. The final step—actually drawing the tree—required very little effort.

Chapters 3–9 will expand considerably on this tutorial, but the majority of your effort will always be devoted to choosing and downloading sequences. MEGA greatly speeds up the download (just click the red cross), but no program can speed up the decision-making process; that *always* requires thought. You need to consider the E value of the hit sequence, the length of the query that aligns with the hit sequence, and—most of all—your knowledge of the biology of the sequences and what you want to accomplish with the phylogeny.

Summary

At this point you should be able to use MEGA 5 to:

- Do a BLAST search to identify a set of subject sequences that are homologous to a sequence of interest (the query sequence).
- Select from that set of sequences those that will be used to create a phylogeny.
- Download those sequences into MEGA's Alignment Explorer.
- Save the downloaded sequences.
- Align the sequences with MUSCLE.
- Make a Neighbor Joining tree from the alignment.
- Display, save, and print that tree.

It would now be valuable to pick a sequence of interest to you and go through the same steps to put that sequence onto a phylogenetic tree. When you finish, move on to Chapter 3.

Acquiring the Sequences

As pointed out in Chapter 2, the most time-consuming part of estimating a phylogenetic tree is acquiring the sequences that will be the tips of the tree. This chapter discusses how to find related sequences and the criteria to consider as you decide which sequences to include. Before you can make those decisions, however, you will need to know what you want to accomplish with the tree. Will it be a "reference" that shows the relationships among all known homologs—at least until next week when another homologous sequence is added to the databases? Do you want it to show the detailed relationships among very closely related sequences, or even different isolates of the same species? Or is it intended to show the deepest possible relationships among the most distantly related sequences? In short, *why* are you making this tree in the first place? After all, a phylogenetic tree is not a decoration to make a paper more attractive to a journal editor, nor is it a means to accommodate a colleague (whose reasons for suggesting a tree may have no apparent application to your own research). A tree is a tool to help better understand a particular biological problem. Only when you are clear about the relationship between the tree and the problem can you decide which sequences to include on the tree.

Once you obtain the sequences you will need to align them. Chapter 4 will discuss the crucial process of sequence alignment.

Hunting Homologs: What Sequences Can Be Included on a Single Tree?

Homology must be distinguished from similarity. *Homology* means that two taxa or sequences are descended from a common ancestor and it implies that, in an alignment, identical residues at a site are identical by descent. *Similarity* merely reflects the proportion of sites that are identical by character state. Two nonhomologous sequences can be aligned and some sites will be identical, but that identity is not the result of descent from a common ancestor. Obviously, it is meaningless to put two nonhomologous sequences onto the same tree, no matter how similar the sequences may be, because the purpose of a phylogenetic tree is to show a process of descent from common ancestors.

Of course, in one sense, it can be said that all sequences are descended from a common ancestral sequence. As genes and proteins evolve, however, they diverge to the point where two genes may share so little sequence that they resemble each other no more than any two randomly chosen sequences. At that point their sequence homology has disappeared and these two sequences should never appear on the same sequence-based tree. However, trees that include nonhomologous sequences are published surprisingly often. It is not uncommon for a set of enzymes that share similar catalytic properties and mechanisms to be given a common designation and subsequently placed on the same tree, regardless of their actual sequence homology.

Becoming More Familiar with BLAST

As discussed in Chapter 2, a BLAST search is the primary tool for identifying sequences that are homologous to your sequence of interest (the query sequence). Although this chapter is not intended as a follow-along tutorial, you may certainly do so if you wish. The file `ebgC.txt` includes both the *ebgC* DNA coding sequence and the corresponding protein sequence that will be used to illustrate this and later chapters. The *ebgC* gene encodes the β subunit of the cryptic enzyme EBG β-galactosidase of *E. coli*.

 Chapter 3: ebgC.txt

Recall from Chapter 2 that you get to the BLAST search page by choosing **Do BLAST search** from the **Align** menu on MEGA's main page (see Figure 2.1). From the BLAST search page click the **Home** button (circled in Figure 2.2) to go to the BLAST main page (**Figure 3.1**), which is divided into three sections.

The top section is for searching the assembled reference sequence genomes. It is valuable if you want to confine your search to sequenced genomes of one of the listed species, or to genomes of microbes.

The middle section is concerned with the search method that you want to employ. If your query sequence is nucleotides and you want to search nucleotide sequences, choose **nucleotide blast** as was illustrated in Chapter 2. Choose **protein blast** if your query is a protein sequence and you want to search protein sequences (which I will describe later in this chapter).

But what if your query sequence is a nucleotide sequence that you *suspect* encodes a protein—but you don't know what protein, or which of six possible reading frames might encode it? You might want to search among the protein sequences for homologs by choosing **blastx** that will translate your nucleotide query in all six frames and search the protein database with each of those translations. The **tblastn** choice works just the opposite of blastx: tblastn uses your protein query to search nucleotide databases, translating them as it goes along. The final choice, **tblastx** translates your nucleotide query in all six frames and uses it to search the nucleotide databases, translating them as it goes along.

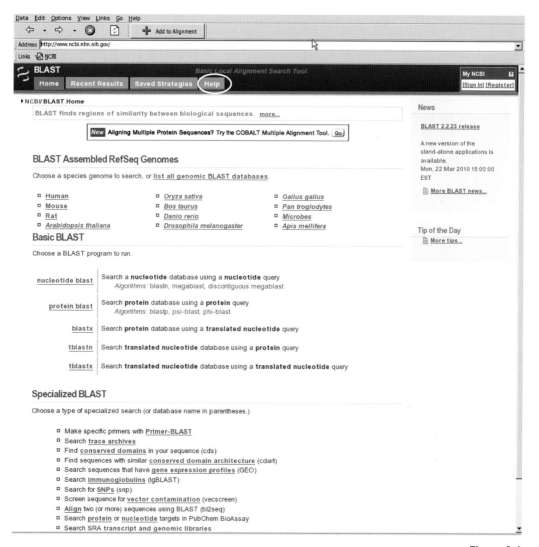

Figure 3.1

As you might imagine, blastx, tblastn, and tblastx searches are considerably slower than blastn or blastp searches. All that translating takes considerable computer effort. But we may wish to go to all that effort because amino acids have 20 possible states, whereas nucleotides have only 4 possible states. As discussed later in this chapter, searching protein sequences can detect much more distantly related homologs than can searching nucleotide sequences.

The bottom section is for specialized searches. Only one of those, the **Align two (or more) sequences using BLAST (bl2seq)** algorithm, will be discussed later in this chapter.

BLAST help

Clicking the **Help** button a the top of the BLAST main page (circled in Figure 3.1) takes you to a page of links to documentation files. Some of the documentation (particularly the **Blast interface description**) is out of date, but most is quite useful. It is worth your while to read through **Blast Program Selection Guide**, **Frequently Asked Questions**, and **The Statistics of Sequence Similarity Score** (which explains both the bit score and the E values quite nicely). The **BLAST glossary** explains the terms that are used and is quite helpful. There is also a **Email blast-help** link that allows you to email questions to the BLAST help desk. They are moderately responsive.

Using the Nucleotide BLAST Page

The Nucleotide BLAST (**blastn**) page (**Figure 3.2**) was discussed in Chapter 2, but two important features were not mentioned. There are five tabs at the top of the page (circled in Figure 3.2). Clicking a tab takes you directly to one of the other BLAST search pages without having to go to the main BLAST page.

Figure 3.2

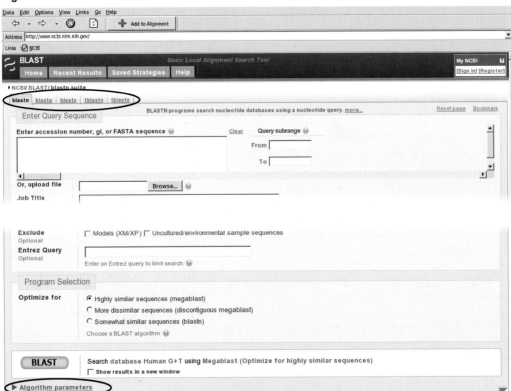

Clicking **Algorithm parameters**, circled near the bottom of Figure 3.2, reveals a region that allows you to modify the BLAST search parameters. Under **General Parameters** the only choice that you are likely to want to modify is **Max Target Sequences**, that controls the maximum number of alignments that will be displayed. The default is 100, but when there are many closely related sequences in the database you may need to increase that number in order to display more distantly related sequences. If you find that your search lists only 100 sequences and that no distantly related sequences (those with E values $>10^{-3}$) are shown, you may need to set a higher limit.

Figure 3.3 shows the result of a BLAST search using the *E. coli* K12 *ebgC* coding sequence as a query in a blastn search. The DNA sequence was pasted into the sequence search box (we could just as easily have entered the accession number and specified the sequence range), the **Nucleotide collection (nr/ nt)** database was chosen, and the program selection was set to **Somewhat similar sequences (blastn)**.

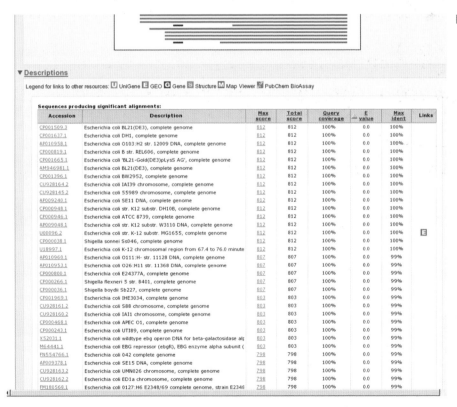

Figure 3.3

▼ **Descriptions**

Legend for links to other resources: U UniGene E GEO G Gene S Structure M Map Viewer PubChem BioAssay

Sequences producing significant alignments:

Accession	Description	Max score	Total score	Query coverage	E value	Max ident	Links
CP001509.3	Escherichia coli BL21(DE3), complete genome	812	812	100%	0.0	100%	
CP001637.1	Escherichia coli DH1, complete genome	812	812	100%	0.0	100%	
AP010958.1	Escherichia coli O103:H2 str. 12009 DNA, complete genome	812	812	100%	0.0	100%	
CP000819.1	Escherichia coli B str. REL606, complete genome	812	812	100%	0.0	100%	
CP001665.1	Escherichia coli 'BL21-Gold(DE3)pLysS AG', complete genome	812	812	100%	0.0	100%	
AM946981.1	Escherichia coli BL21(DE3), complete genome	812	812	100%	0.0	100%	
CP001396.1	Escherichia coli BW2952, complete genome	812	812	100%	0.0	100%	
CU928164.2	Escherichia coli IAI39 chromosome, complete genome	812	812	100%	0.0	100%	
CU928145.2	Escherichia coli 55989 chromosome, complete genome	812	812	100%	0.0	100%	
AP009240.1	Escherichia coli SE11 DNA, complete genome	812	812	100%	0.0	100%	
CP000948.1	Escherichia coli str. K12 substr. DH10B, complete genome	812	812	100%	0.0	100%	
CP000946.1	Escherichia coli ATCC 8739, complete genome	812	812	100%	0.0	100%	
AP009048.1	Escherichia coli str. K12 substr. W3110 DNA, complete genome	812	812	100%	0.0	100%	
U00096.2	Escherichia coli str. K-12 substr. MG1655, complete genome	812	812	100%	0.0	100%	E
CP000038.1	Shigella sonnei Ss046, complete genome	812	812	100%	0.0	100%	
U18997.1	Escherichia coli K-12 chromosomal region from 67.4 to 76.0 minute	812	812	100%	0.0	100%	
AP010960.1	Escherichia coli O111:H- str. 11128 DNA, complete genome	807	807	100%	0.0	99%	
AP010953.1	Escherichia coli O26:H11 str. 11368 DNA, complete genome	807	807	100%	0.0	99%	
CP000800.1	Escherichia coli E24377A, complete genome	807	807	100%	0.0	99%	
CP000266.1	Shigella flexneri 5 str. 8401, complete genome	807	807	100%	0.0	99%	
CP000036.1	Shigella boydii Sb227, complete genome	807	807	100%	0.0	99%	
CP001969.1	Escherichia coli IHE3034, complete genome	803	803	100%	0.0	99%	
CU928161.2	Escherichia coli S88 chromosome, complete genome	803	803	100%	0.0	99%	
CU928160.2	Escherichia coli IAI1 chromosome, complete genome	803	803	100%	0.0	99%	
CP000468.1	Escherichia coli APEC O1, complete genome	803	803	100%	0.0	99%	
CP000243.1	Escherichia coli UTI89, complete genome	803	803	100%	0.0	99%	
X52031.1	Escherichia coli wildtype ebg operon DNA for beta-galactosidase alp	803	803	100%	0.0	99%	
M64441.1	Escherichia coli EBG repressor (ebgR), EBG enzyme alpha subunit (	803	803	100%	0.0	99%	
FN554766.1	Escherichia coli 042 complete genome	798	798	100%	0.0	99%	
AP009378.1	Escherichia coli SE15 DNA, complete genome	798	798	100%	0.0	99%	
CU928163.2	Escherichia coli UMN026 chromosome, complete genome	798	798	100%	0.0	99%	
CU928162.2	Escherichia coli ED1a chromosome, complete genome	798	798	100%	0.0	99%	
FM180568.1	Escherichia coli O127:H6 E2348/69 complete genome, strain E2348	798	798	100%	0.0	99%	

There are 71 hits, but the first 16 or so are identical to the query, and the next 30 or so differ from it by <2% identity. Those are all either *E. coli* or *Shigella*, which is so closely related to *E. coli* that, except for medical purposes, it can be considered the same species. Of the remaining sequences (**Figure 3.4**), only 5 meet the criteria of aligning over >60% of the query length with an E value <10^{-3}. If we are interested in looking at the relationships of the *ebgC* gene from different species, we are left with a tree of six sequences, four of which are from the genus *Vibrio*.

Figure 3.4

CP001063.1	Shigella boydii CDC 3083-94, complete genome	450	450	55%	4e-123	100%	
FP929040.1	Enterobacter cloacae subsp. cloacae NCTC 9394 draft genome	260	260	95%	4e-66	73%	
CP000789.1	Vibrio harveyi ATCC BAA-1116 chromosome I, complete sequence	114	114	88%	3e-22	67%	
CP001805.1	Vibrio sp. Ex25 chromosome 1, complete sequence	111	111	88%	4e-21	66%	
BA000031.2	Vibrio parahaemolyticus RIMD 2210633 DNA, chromosome 1, comp	96.9	96.9	88%	9e-17	65%	
AE016795.2	Vibrio vulnificus CMCP6 chromosome I complete sequence	93.3	93.3	61%	1e-15	68%	
BA000037.2	Vibrio vulnificus YJ016 DNA, chromosome I, complete sequence	82.4	82.4	39%	2e-12	70%	
CP000462.1	Aeromonas hydrophila subsp. hydrophila ATCC 7966, complete ger	60.8	60.8	38%	7e-06	67%	
CP001133.1	Vibrio fischeri MJ11 chromosome II, complete sequence	57.2	57.2	11%	8e-05	84%	
CP000021.2	Vibrio fischeri ES114 chromosome II, complete sequence	51.8	51.8	11%	0.003	82%	
XM_002429891.1	Pediculus humanus corporis sodium/potassium/calcium exchanger	42.8	42.8	8%	1.8	86%	G
CP001874.1	Thermobispora bispora DSM 43833, complete genome	41.0	41.0	6%	6.3	92%	
EU138787.1	Bacillus subtilis strain NRRL BD-617 phosphoribosylaminoimidazolec	41.0	41.0	4%	6.3	100%	
EU138778.1	Bacillus subtilis strain NRRL BD-594 phosphoribosylaminoimidazolec	41.0	41.0	4%	6.3	100%	
EU138759.1	Bacillus subtilis strain NRRL B-41008 phosphoribosylaminoimidazole	41.0	41.0	4%	6.3	100%	
EU138756.1	Bacillus subtilis strain NRRL B-23974 phosphoribosylaminoimidazole	41.0	41.0	4%	6.3	100%	
EU138749.1	Bacillus sonorensis strain NRRL B-23154 phosphoribosylaminoimida	41.0	41.0	4%	6.3	100%	
EU138741.1	Bacillus subtilis subsp. spizizenii strain NRRL B-23050 phosphoribos	41.0	41.0	4%	6.3	100%	
BC157305.1	Xenopus tropicalis vitronectin, mRNA (cDNA clone IMAGE:7612281)	41.0	41.0	6%	6.3	90%	U G
AC034232.8	Homo sapiens chromosome 5 clone CTD-2233C11, complete seque	41.0	41.0	4%	6.3	100%	
AC016647.7	Homo sapiens chromosome 16 clone RP11-89G14, complete seque	41.0	41.0	7%	6.3	88%	

It is certainly possible that *ebgC* is in fact this narrowly distributed, but we would really like to know whether homologs exist in more distantly related species. The problem we face in finding more distantly related homologs is the result of a *lower limit of detectable homology*. For DNA sequences, there are only four possible states (A, C, G, or T) of each character, so when sequences have diverged sufficiently that they remain identical at approximately 25% of their sites, the sequences appear to be no more closely related than any two random, nonhomologous sequences. The solution to finding sequences that are more distantly related is to search using a protein sequence as the query.

Using BLAST to Search for Related Protein Sequences

Proteins have 20 possible states at each site rather than four, so the lower limit of detectable homology drops to about 5% similarity (rather than the 25% threshold for nucleotides). The BLAST Protein (**blastp**) page allows you to use selected protein sequences as the basis for a search for related protein sequences.

Figure 3.5 shows a BLAST search using the EbgC protein sequence as a query. From the **blastn** page, I clicked the **blastp** tab (see Figure 3.2). I pasted the protein accession number `NP_417548` into the search box, chose the default **Non-redundant protein sequence (nr)** database, and specified the default **blastp (protein-protein BLAST)** program. (*Protein-protein* means that the query is a protein and protein sequences are being searched.) The resulting screen is seen in Figure 3.5.

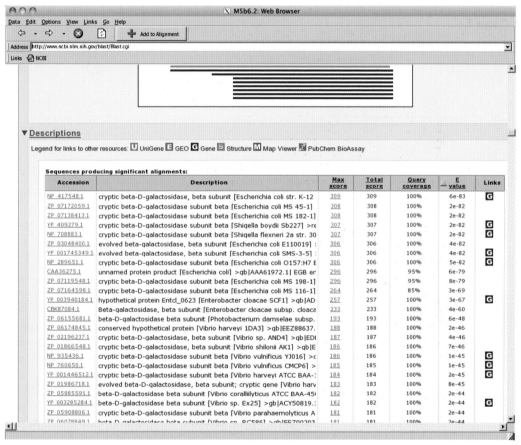

Figure 3.5

Notice that the overall appearance of the window in Figure 3.5 is slightly different from that in Figure 3.3, the main difference being that the **blastp** window does not show a **Max Ident** column. The protein–protein BLAST result identified 100 hits, of which 58 sequences have E values <0.001. All but two of those sequences have query coverage >60%, so clearly there has been some gain by using a protein query.

The first alignment appears very strange (**Figure 3.6**). Instead of one or two alignments being shown, there are links to no fewer than 115 different files (click the **113 more sequence titles** link, indicated by a red arrow in Figure 3.6) from this one "hit" alone! The alignment heading states that these are identical to the query at 149/149 sites (i.e., they are perfect matches). As expected, the first hit is the query sequence itself. Among the files listed are several *E. coli* strains and one *Shigella* strain. Which of these files should be included on the tree? Perhaps all of them; it depends entirely on the purpose of the tree.

Figure 3.6

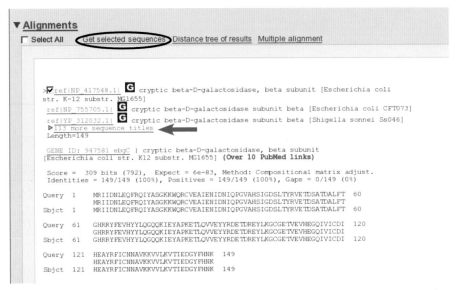

Although the protein sequences are identical to the query, and thus to each other, the nucleotide sequences of the genes encoding these proteins may differ from each other because of **silent substitutions**—mutations that do not change the amino acid being specified. If you are interested in looking at the finest structure of the tree (i.e., structure that distinguishes *all* existing differences), you will want to include all the different DNA sequences, even those that encode identical proteins. For purposes of this chapter, however, we are not interested in the finest possible structure, but in estimating the relationships among *ebgC* genes from different species.

We can scan down the alignments, ticking the check boxes of the sequences we are interested in. Those will certainly include the first sequence. I'll choose one *Shigella* sequence and one from a pathogenic *E. coli* O157:H7 strain. After that I'll pick one from each species for which E <10⁻³ and the query coverage is at least 60%. Finally, I'll scroll back to the beginning of the alignments on the page and click the **Get selected sequences** link (circled in Figure 3.6) to bring up a list of the *protein* sequence files we have selected (**Figure 3.7**).

Figure 3.7

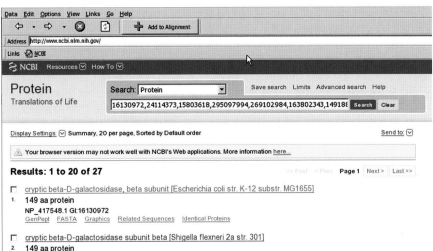

Clicking the first link takes us to the protein sequence file of EbgC from *E. coli* strain K12 substrain MG1655 (**Figure 3.8**).

Figure 3.8

However, we do *not* want to put that protein sequence into the Alignment Explorer for two reasons. First, basing the tree on protein sequences means we will lose the ability to distinguish sequences that differ only by silent substitutions. Second, it is generally preferable to base trees on the DNA coding sequences if for no other reason than that some phylogenetic methods, particularly Bayesian Inference and Maximum Likelihood, can be excruciatingly slow when working with protein sequences. So, now that we've chosen more subject sequences based on proteins, it's time to take those subjects back to nucleotide sequences.

Finalizing Selected Sequences for a Tree

Scrolling to the bottom of the screen in Figure 3.8 reveals a **CDS** (coding sequence) link; this link is circled in **Figure 3.9**.

Figure 3.9

```
REFERENCE   15 (residues 1 to 149)
  AUTHORS   Blattner,F.R. and Plunkett,G. III.
    TITLE   Direct Submission
  JOURNAL   Submitted (16-JAN-1997) Laboratory of Genetics, University of
            Wisconsin, 425G Henry Mall, Madison, WI 53706-1580, USA
  COMMENT   PROVISIONAL REFSEQ: This record has not yet been subject to final
            NCBI review. The reference sequence is identical to AAC76112.
            Method: conceptual translation.
 FEATURES            Location/Qualifiers
     source          1..149
                     /organism="Escherichia coli str. K-12 substr. MG1655"
                     /strain="K-12"
                     /sub_strain="MG1655"
                     /db_xref="taxon:511145"
     Protein         1..149
                     /product="cryptic beta-D-galactosidase, beta subunit"
                     /function="enzyme; Degradation of small molecules: Carbon
                     compounds"
                     /calculated_mol_wt=17326
     Region          1..149
                     /region_name="DUF386"
                     /note="Domain of unknown function (DUF386); cl01047"
                     /db_xref="CDD:174505"
     CDS             1..149
                     /gene="ebgC"
                     /locus_tag="b3077"
                     /gene_synonym="ECK3067; JW3048"
                     /coded_by="NC_000913.2:3223744..3224193"
                     /GO_process="GO:0016052 - carbohydrate catabolic process"
                     /note="evolved beta-D-galactosidase, beta subunit; cryptic
                     gene"
                     /transl_table=11
                     /db_xref="ASAP:ABE-0010106"
                     /db_xref="UniProtKB/Swiss-Prot:P0AC73"
                     /db_xref="ECOCYC:EG10253"
                     /db_xref="EcoGene:EG10253"
                     /db_xref="GeneID:947581"
 ORIGIN
 //
```

Clicking the **CDS** link takes us to the file that contains the coding sequence (**Figure 3.10**). Notice that the beige **Change Region Shown** area shows the region that encodes this protein. Most, but not all, of the CDS files show the sequence in the correct orientation.

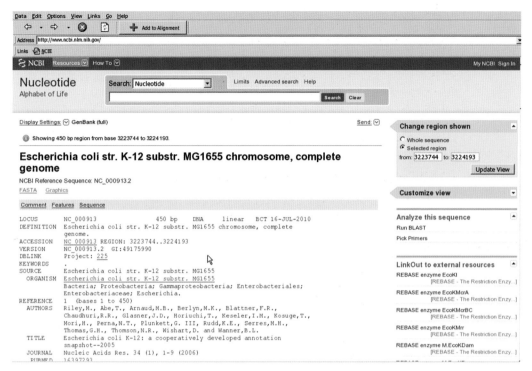

Figure 3.10

Scroll down and look at the coding sequence (**Figure 3.11**). If the description of the CDS that appears just above the sequence itself indicates the "complement" (circled), the sequence is in the wrong orientation. Notice that the beige **Change Region Shown** area shows the region that encodes this protein.

You can easily fix the orientation of the sequence. In the **Customize View** region, choose **Show reverse complement** and click the **Update View** button. Clicking the red cross near the top of the window now adds the sequence to the Alignment Explorer. If you do inadvertently add a sequence that is in the wrong orientation to the Alignment Editor, simply right-click (Macintosh = Control-click) the name of the sequence in the Alignment Editor and choose **Reverse complement** from the list of choices. Likewise, if you accidentally add the protein sequence instead of the DNA coding sequence, right-click the name of the sequence in the Alignment Editor and choose **Delete** from the list of choices.

Figure 3.11

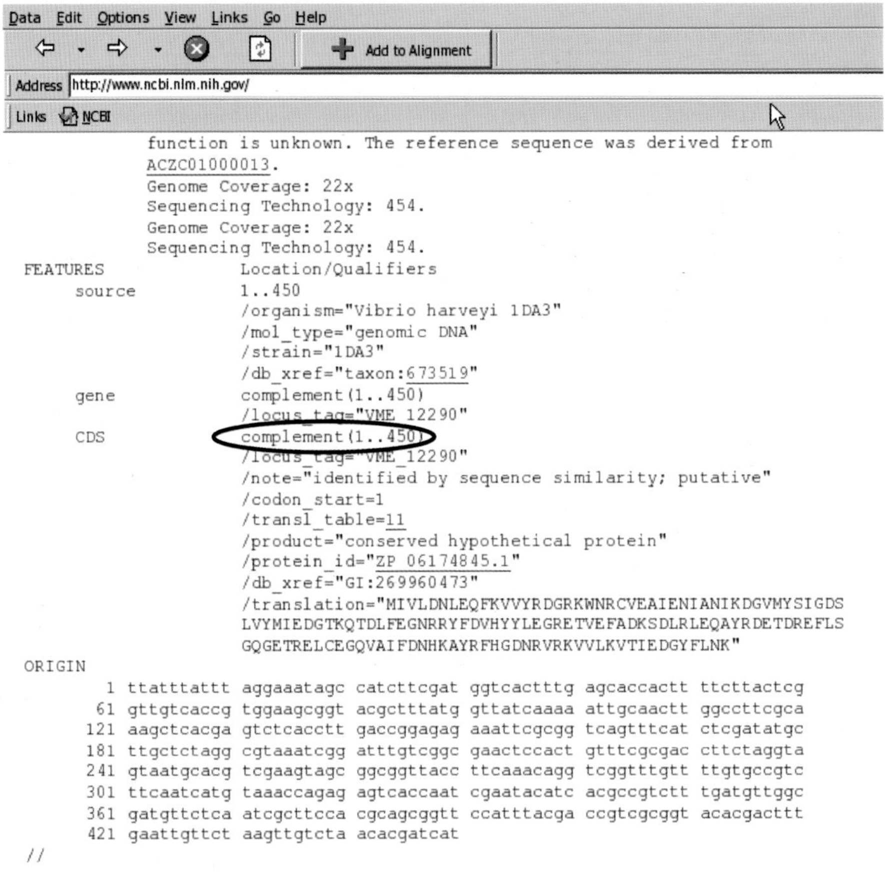


```
                  function is unknown. The reference sequence was derived from
                  ACZC01000013.
                  Genome Coverage: 22x
                  Sequencing Technology: 454.
                  Genome Coverage: 22x
                  Sequencing Technology: 454.
FEATURES                   Location/Qualifiers
     source                1..450
                           /organism="Vibrio harveyi 1DA3"
                           /mol_type="genomic DNA"
                           /strain="1DA3"
                           /db_xref="taxon:673519"
     gene                  complement(1..450)
                           /locus_tag="VME_12290"
     CDS                   complement(1..450)
                           /locus_tag="VME_12290"
                           /note="identified by sequence similarity; putative"
                           /codon_start=1
                           /transl_table=11
                           /product="conserved hypothetical protein"
                           /protein_id="ZP_06174845.1"
                           /db_xref="GI:269960473"
                           /translation="MIVLDNLEQFKVVYRDGRKWNRCVEAIENIANIKDGVMYSIGDS
                           LVYMIEDGTKQTDLFEGNRRYFDVHYYLEGRETVEFADKSDLRLEQAYRDETDREFLS
                           GQGETRELCEGQVAIFDNHKAYRFHGDNRVRKVVLKVTIEDGYFLNK"
ORIGIN
        1 ttatttattt aggaaatagc catcttcgat ggtcactttg agcaccactt ttcttactcg
       61 gttgtcaccg tggaagcggt acgctttatg gttatcaaaa attgcaactt ggccttcgca
      121 aagctcacga gtctcacctt gaccggagag aaattcgcgg tcagtttcat ctcgatatgc
      181 ttgctctagg cgtaaatcgg atttgtcggc gaactccact gtttcgcgac cttctaggta
      241 gtaatgcacg tcgaagtagc ggcggttacc ttcaaacagg tcggtttgtt ttgtgccgtc
      301 ttcaatcatg taaaccagag agtcaccaat cgaatacatc acgccgtctt tgatgttggc
      361 gatgttctca atcgcttcca cgcagcggtt ccatttacga ccgtcgcggt acacgacttt
      421 gaattgttct aagttgtcta acacgatcat
//
```

As discussed in Chapter 2, use the browser's back arrows to get back to the list of protein sequences. The sequences represent quite a range of species, from a group of closely related *E. coli/Shigella* Gram negative sequences to quite distantly related *Streptococcus pneumoniae* Gram positive sequences. I continued adding all the coding (nucleotide) sequences to the Alignment Explorer for a total of 29 sequences (**Figure 3.12**)—many more than the 6 that were obtained using the coding sequence as a query.

The name display area in Figure 3.12 has been widened by dragging the handle (circled) to show the entire sequence description, some of which are very long. By default, MEGA uses up to the first 200 characters of the sequence description as the name. Names of this length pose several problems. First, these names will be used to identify the sequences on the tree that MEGA eventually draws, and long names simply take up too much room. Second,

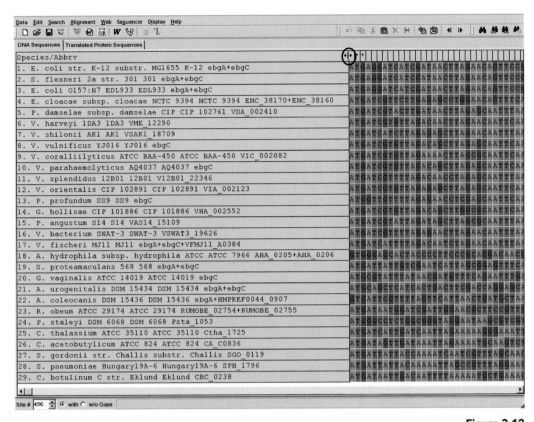

Figure 3.12

the names are often redundant and confusing (notice **E. coli str. K12 substr. MG1655 K-12 ebgA+ebgC** in the first sequence). Finally, some phylogenetics program file formats require names ≤10 characters, while others recognize only the first 30 characters as meaningful. Beyond that, some file formats (notably Nexus) do not permit spaces or characters other than letters, digits, and the underscore character in file names. Characters such as - , \ , ( ), *, and / may cause programs that use the Nexus format to misbehave.* All of these reasons necessitate some editing of the sequence names. Since you might eventually want to export the alignment for use with another program, it is wise to edit the sequence names down to reasonable lengths and to eliminate non-alphanumeric characters.

To edit a sequence name in the Alignment Editor, double-click the name and enter a revised name. It is important to edit the name at this stage because the name cannot easily be edited in the .meg file that MEGA uses for all of its analyses.

---

*Multiple file formats and their quirks constitute one of the major frustrations of phylogenetics. See Appendix I for discussion of file formats.

Some guidelines for sequence names:

- Each name must be unique. No program can accept multiple sequences with the same name.

- Eliminate spaces, replacing them with the underscore (_) character. Many programs will not accept spaces in names.

- Use only letters, numbers, and the underscore character in sequence names. In particular, be sure to eliminate the colon character (:) and the dash character; it is easy to miss (e.g., E. coli K-12).

- Make the names meaningful. Your lab may refer to a particular strain of *C. elegans* simply as WRM22 but that won't be meaningful to others. On the other hand, C_elegans_WRM22 would be fine if you need to distinguish that particular strain from other *C. elegans* strains.

Figure 3.13 shows the Alignment Editor after editing the names. Notice that I failed to eliminate a forbidden character in one sequence. Can you find the error?

Figure 3.13

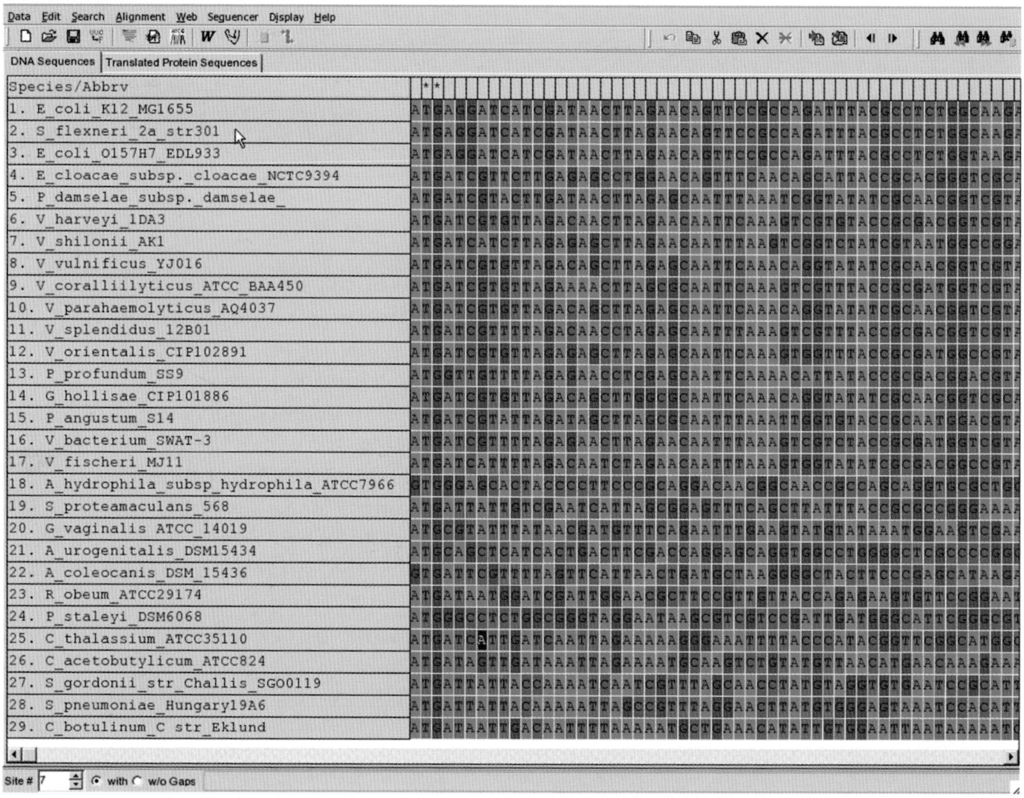

 Chapter 3: ebgC.mas

At this point I saved the alignment as `ebgC.mas` as described in Chapter 2. To save time, if you wish you can download that alignment file. Remember, however, that these sequences are not yet aligned; they are simply saved in the Alignment Editor. Chapter 4 will describes the next steps in preparing your data to make a valid tree.

Other Ways to Find Sequences of Interest (Beware! The Risks Are High)

We may wonder if there are additional *ebgC* genes we missed with our BLAST search. We can query the databases by searching on the word `ebgC`. In the main MEGA window, choose **Query Databanks** from the **Alignment** window. Again, MEGA opens a browser window to the NCBI website (**Figure 3.14**). Change **Nucleotide** (circled) to **Protein**, enter `ebgC` in the search box, then click the **Search** button.

Figure 3.14

The search retrieves 261 files (**Figure 3.15**). To display all of those on just two web pages click the **Display Settings** link (circled in Figure 3.15) and change **Items per page** to 200.

Figure 3.15

Scrolling down the page we see that most of the sequences are about 149 amino acids, but as we get down to sequences 156–161 we see a series of proteins from *Bacteroides* species in the range of 201–208 amino acids. Those proteins are called EbgC, so we could just click the **CDS** links and add them to the Alignment Editor, right? **No, no, no! Danger!** Those sequences were not recovered in the BLAST search, so we need to ask, *why* did BLAST not find this particular EbgC? Was the E value too high? Did the sequences align over too short a portion of the query? One very real possibility is that the *Bacteroides* EbgC proteins are not homologous to *E. coli* EbgC.

We can test the homology between the EbgC proteins of *E. coli* and those of *Bacteroides* by using the **Blast 2 Sequences** tool. Open a new browser window by choosing **New Window** from the **Data** menu, then enter http://www.ncbi. nlm.nih.gov/blast/ in the address window and hit the **Enter** key to go to the BLAST main page. (You can save the link to this page by choosing **Add** from the **Links** menu of the browser window.) In the **Specialized BLAST** section at the bottom of the page click the **Align two (or more) sequences** link. This will open the **Blast 2 Sequences (bl2seq)** window (**Figure 3.16**).

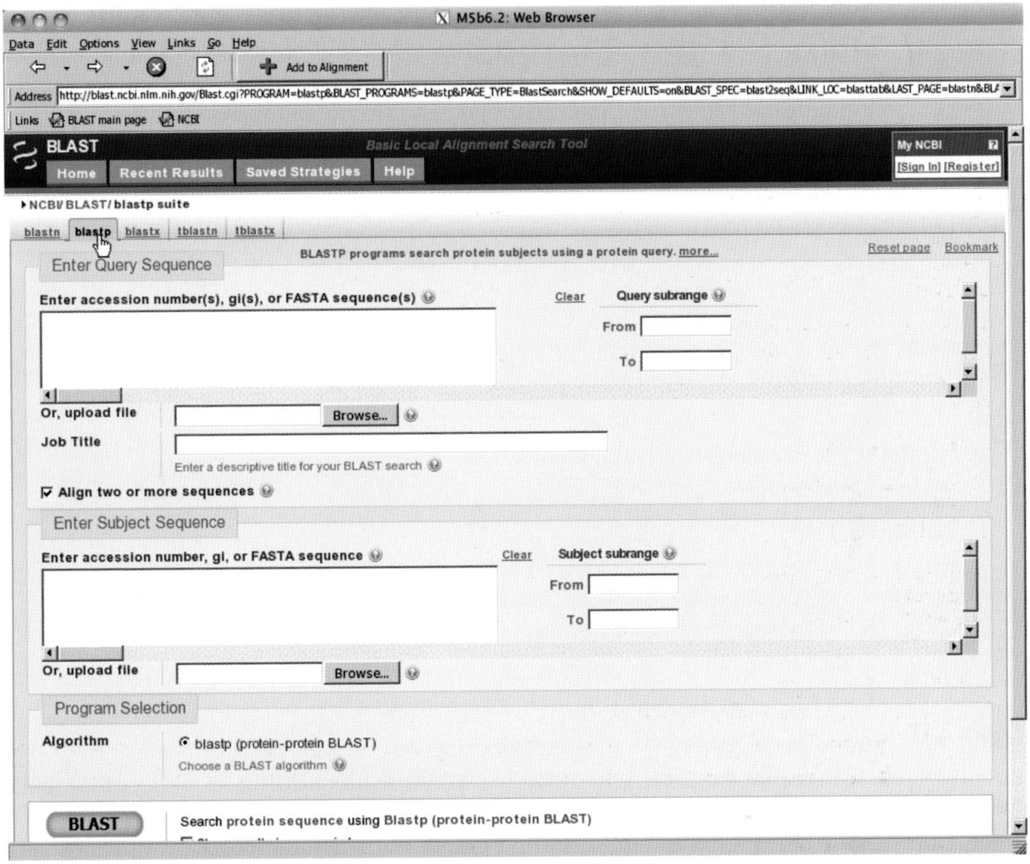

Figure 3.16

Paste the *E. coli* EbgC protein accession number, NP_417548, into the upper search box, and enter the accession number of one of the *Bacteroides* protein sequence (ZP_07038536.1, for instance) into the lower search box. Then change the program from **blastn** to **blastp** by clicking the tab indicated by the hand cursor and click the **BLAST** button located below the lower search box.

The result shows a query coverage of only 40% and an E value of 0.16 (**Figure 3.17**). Running **bl2seq** on the other *Bacteroides* sequences gives similar results. Thus, despite having the same name, the *Bacteroides* EbgC sequences are not homologs of the *E. coli* EbgC sequence and do not belong on the tree.

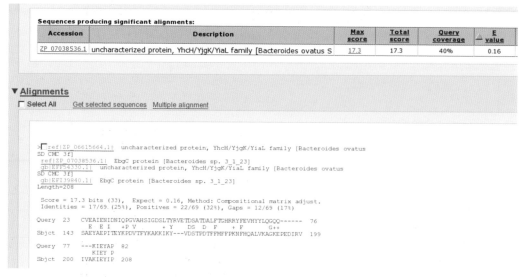

Figure 3.17

There may be times when you want to consider adding sequences that were not found by a BLAST search. Perhaps a colleague sends you a sequence. Before adding such a sequence, however *always* use the **bl2seq** tool to test for homology. Even if you find homology, be sure that the E value is within your acceptable range and check the length over which the two sequences align. Later, in Chapter 12, I will consider an alternative way to assess whether a particular sequence belongs in an alignment.

CHAPTER **4**

Aligning the Sequences

MEGA 5 offers two methods for aligning nucleotide and protein sequences: ClustalW (Thompson et al. 1994) and MUSCLE (MUltiple Sequence Comparison by Log-Expectation; Edgar 2004b). Although ClustalW is more widely used, MUSCLE is slightly more accurate (Nuin et al. 2006) and is 2–5 times faster in typical-size data sets. The speed of modern desktop computers makes speed a non-issue for most data sets; for example, for the SmallData data set, MUSCLE required 3 seconds on my computer while ClustalW required 6 seconds. On the other hand, I recently had a large data set that ClustalW required several hours to align; MUSCLE aligned the same data set in a few minutes. Indeed, the major advantage of MUSCLE over ClustalW is its handling of large data sets. For a set of 5000 sequences of average length 350, MUSCLE was over 80,000 times faster than ClustalW (Edgar 2004b). The continuously accelerating rate of DNA sequence accumulation in the databases, and especially that of genome sequence accumulation, suggests that the need to work with very large data sets will become increasingly common.

Aligning Sequences with MUSCLE

Download Chapter 3:ebgC.mas

The 29 *ebgC* sequences that were downloaded to MEGA's Alignment Editor in Chapter 3 were saved as `ebgC.mas`. After opening the file, notice the **Translated Protein Sequences** tab (red arrow in **Figure 4.1**). Clicking that tab translates the coding sequences into the corresponding protein sequences. Figure 4.1 shows the C-termini of the proteins. Note that the final character is always an asterisk (*), which corresponds to the chain termination codon; obviously, no asterisks appear within the sequences.

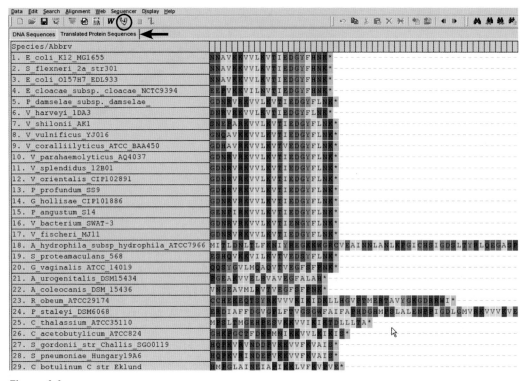

Figure 4.1

Click the **DNA Sequences** tab to return to the DNA view (see Figure 3.13), where we can align the sequences with MUSCLE (as discussed in Chapter 2) by choosing **Align by Muscle** from the **Alignment** menu, or by clicking the "muscle" icon in the toolbar (circled in Figure 4.1). Choose **Align DNA** and accept the default parameters in the **Muscle Alignment Parameters** window by clicking the **Compute** button.

When we translate the resulting DNA alignment to protein sequences, however, we get sequences that are replete with question marks and asterisks (**Figure 4.2**)—the latter signifying termination codons *within* the sequence! What happened?

MUSCLE introduced gaps according to its algorithm. When gaps appeared within codons, the translation program encountered an unidentifiable codon and indicated that with a ?. When single or double gaps were introduced, they shifted the reading frame, resulting in a nonsense codon downstream. Gaps are intended to indicate historical insertions or deletions (indels). If those indels had actually occurred in ancestral proteins, the proteins would have been inactivated and no descendants would be available to us. Clearly, MUSCLE misplaced the gaps into biologically unrealistic positions.

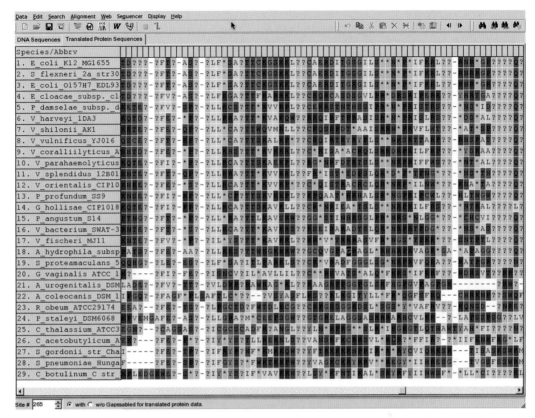

Figure 4.2

Why would the program do such an obviously silly thing? Because MUS-CLE knows nothing of biology or of the functional constraints imposed by frame shifts; it simply attempts to maximize an alignment score. We can solve the problem of misplaced gaps (and thus misaligned bases) by aligning the protein sequences instead of aligning the DNA sequences.

Click the **DNA Sequences** tab to return to the nucleotide view, then type Control-A to select all the sequences. From the **Edit** menu choose **Delete Gaps**. We are now back where we started, with unaligned DNA sequences (see Figure 3.13). Click the **Align by Muscle** icon and select **Align Codons** (**Figure 4.3**). When the **Muscle Parameters** window appears, just click the **Compute** button as before.

Figure 4.3

Figure 4.4 shows the translated protein sequences, and this time there are no ? or * within the aligned protein sequences. Because MUSCLE treated each codon as a unit, gaps were introduced only between codons.

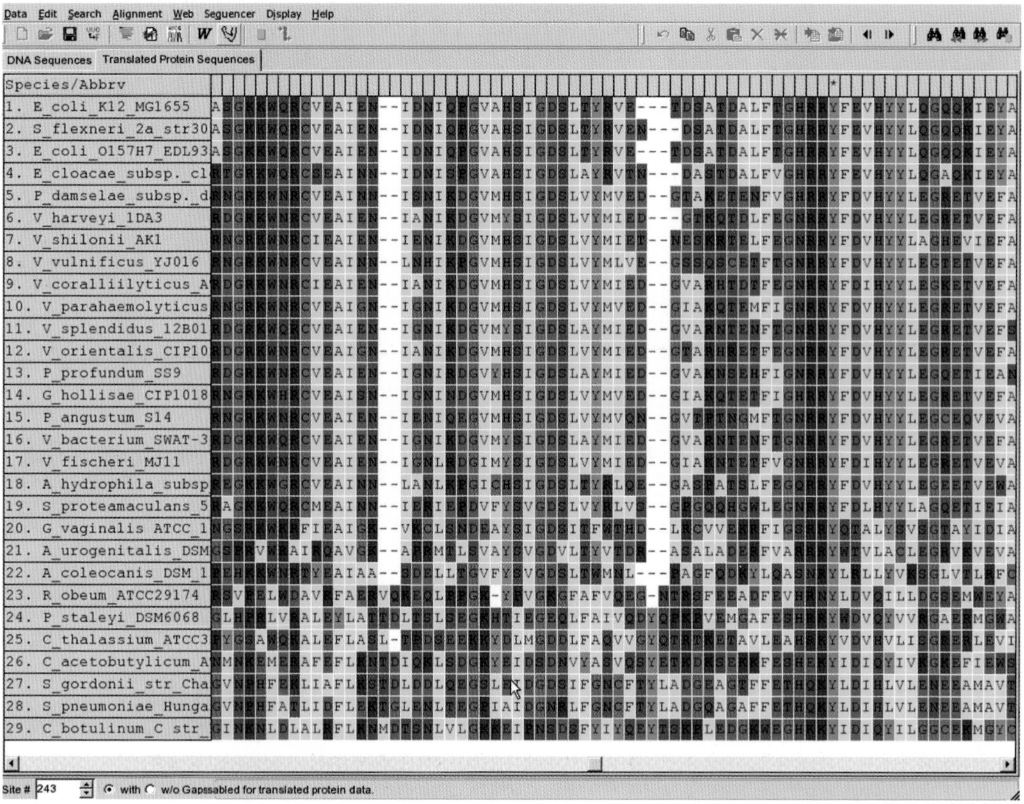

Figure 4.4

Notice that in the DNA sequence view gaps are all multiples of three (**Figure 4.5**). Alignments of coding sequences are much more accurate when aligned by codons than when directly aligned by the nucleotide sequences (Hall 2005). **Always align coding sequences by codons.**

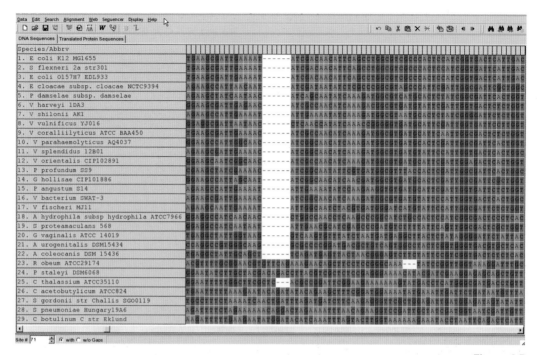

Figure 4.5

Examine and Possibly Manually Adjust the Alignment

You can and should examine the alignment by eye to see if there are any obviously misaligned sites. The key word here is "obviously." Don't be tempted to fool excessively with the alignment; in general, both the MUSCLE and the ClustalW algorithms are excellent and it is not likely that modifications will improve matters.

Trim excess sequence

Sometimes you may encounter a sequence that is much longer than the majority but is not obviously "wrong." Such a sequence may be the result of the fusion of two genes that more typically encode different subunits. In such a case you may remove the excess sequence by selecting the excess and typing `Control-X` to delete it. I selected the first base in sequence #1 of **Figure 4.6**.

The **Site #** box at the bottom shows that that site is at position 385 in the alignment. If we scroll to the left we find that the first base of sequence #24 is at position 298, and the first base of sequences #18 is the only one at position #1 of the alignment. Those two sequences have long 5′ tail segments that do not align with the rest of the sequences. Those tails will contribute nothing

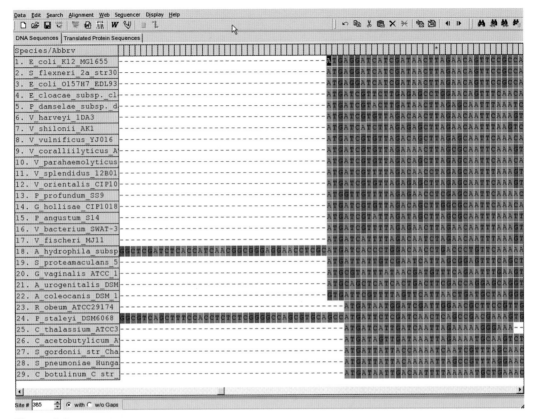

Figure 4.6

to the alignment. We can select everything from position #1 through position #381 by dragging (**Figure 4.7A**), then we can delete the selected region by typing `Control-X` (**Figure 4.7B**). Sequence #24 also has a long tail at the 3′ end that could be eliminated in the same way.

Save the alignment as `ebgC_aligned.mas`. The alignment should now be exported as a MEGA file by choosing **Export Alignment** then **Mega format** from the **Data** menu. This alignment was saved as `ebgC.meg`. When an alignment is complete, always save it as a .mas file even though you will also save it as a .meg file. See Appendix III for a discussion of the importance of always saving a .mas file.

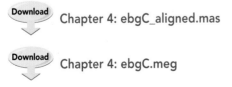

Chapter 4: ebgC_aligned.mas

Chapter 4: ebgC.meg

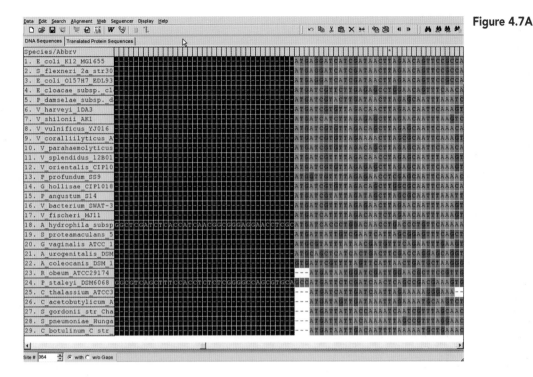

Figure 4.7A

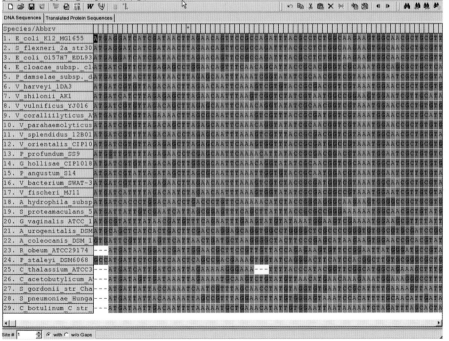

Figure 4.7B

Eliminate duplicate sequences

Although we tried not to do so, we may have included some duplicate sequences. Duplicate sequences add no information to the tree, increase the computation time, and clutter the appearance of the tree, so they should be eliminated.

In the main MEGA window choose **Open a File/Session** from the **File** menu to open the `ebgC.meg` file (**Figure 4.8A**), then choose **Compute Pairwise Distances** from the **Distances** menu (**Figure 4.8B**).

Figure 4.8A

Figure 4.8B

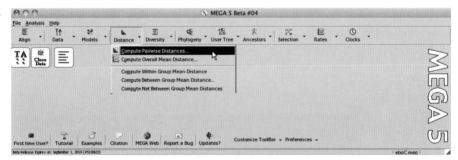

When the **Analysis Preferences** window opens (**Figure 4.9**), click in the yellow area at the right end of the line that says **Model/Method** and change the model to **No. of differences**. Then click the **Compute** button.

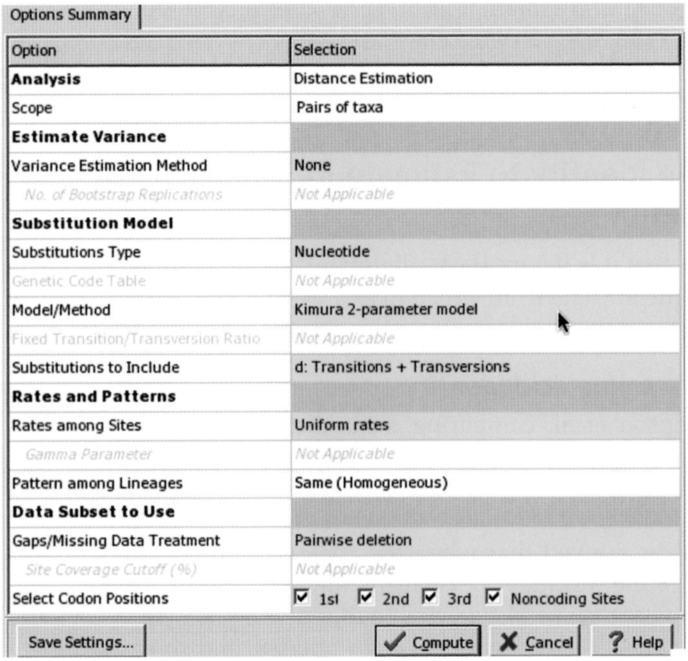

Figure 4.9

Options Summary

Option	Selection
Analysis	Distance Estimation
Scope	Pairs of taxa
Estimate Variance	
Variance Estimation Method	None
No. of Bootstrap Replications	*Not Applicable*
Substitution Model	
Substitutions Type	Nucleotide
Genetic Code Table	*Not Applicable*
Model/Method	Kimura 2-parameter model
Fixed Transition/Transversion Ratio	*Not Applicable*
Substitutions to Include	d: Transitions + Transversions
Rates and Patterns	
Rates among Sites	Uniform rates
Gamma Parameter	*Not Applicable*
Pattern among Lineages	Same (Homogeneous)
Data Subset to Use	
Gaps/Missing Data Treatment	Pairwise deletion
Site Coverage Cutoff (%)	*Not Applicable*
Select Codon Positions	☑ 1st ☑ 2nd ☑ 3rd ☑ Noncoding Sites

Save Settings... ✓ Compute ✗ Cancel ? Help

A results window showing pairwise distances opens (**Figure 4.10**). You will need to enlarge the window to see the entire matrix of pairwise distances. The distances shown are the number of differences between the two sequences being compared. When that distance is zero, the sequences are identical. To make it easier to pick out the zeros, click the down-pointing arrow near the upper right of the window (circled) to reduce the number of decimal places shown to zero.

Figure 4.10

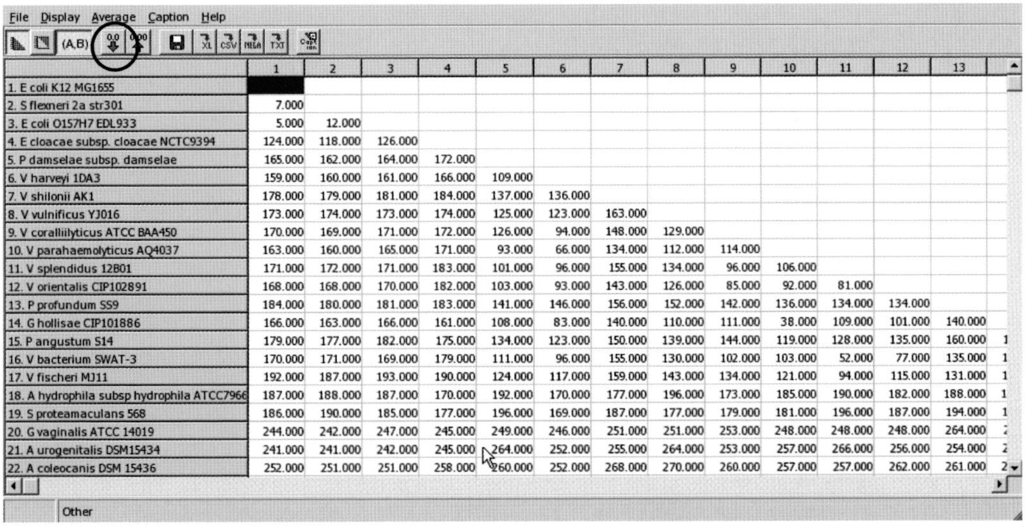

File Display Average Caption Help

	1	2	3	4	5	6	7	8	9	10	11	12	13	
1. E coli K12 MG1655														
2. S flexneri 2a str301	7.000													
3. E coli O157H7 EDL933	5.000	12.000												
4. E cloacae subsp. cloacae NCTC9394	124.000	118.000	126.000											
5. P damselae subsp. damselae	165.000	162.000	164.000	172.000										
6. V harveyi 1DA3	159.000	160.000	161.000	166.000	109.000									
7. V shilonii AK1	178.000	179.000	181.000	184.000	137.000	136.000								
8. V vulnificus YJ016	173.000	174.000	173.000	174.000	125.000	123.000	163.000							
9. V coralliilyticus ATCC BAA450	170.000	169.000	171.000	172.000	126.000	94.000	148.000	129.000						
10. V parahaemolyticus AQ4037	163.000	160.000	165.000	171.000	93.000	66.000	134.000	112.000	114.000					
11. V splendidus 12B01	171.000	172.000	171.000	183.000	101.000	96.000	155.000	134.000	96.000	106.000				
12. V orientalis CIP102891	168.000	168.000	170.000	182.000	103.000	93.000	143.000	126.000	85.000	92.000	81.000			
13. P profundum SS9	184.000	180.000	181.000	183.000	141.000	146.000	156.000	152.000	142.000	136.000	134.000	134.000		
14. G hollisae CIP101886	166.000	163.000	166.000	161.000	108.000	83.000	140.000	110.000	111.000	38.000	109.000	101.000	140.000	
15. P angustum S14	179.000	177.000	182.000	175.000	134.000	123.000	150.000	139.000	144.000	119.000	128.000	135.000	160.000	1
16. V bacterium SWAT-3	170.000	171.000	169.000	179.000	111.000	96.000	155.000	130.000	102.000	103.000	52.000	77.000	135.000	1
17. V fischeri MJ11	192.000	187.000	193.000	190.000	124.000	117.000	159.000	143.000	134.000	121.000	94.000	115.000	131.000	1
18. A hydrophila subsp hydrophila ATCC7966	187.000	188.000	187.000	170.000	192.000	170.000	177.000	196.000	173.000	185.000	190.000	182.000	188.000	1
19. S proteamaculans 568	186.000	190.000	185.000	177.000	196.000	169.000	187.000	177.000	179.000	181.000	196.000	187.000	194.000	1
20. G vaginalis ATCC 14019	244.000	242.000	247.000	245.000	249.000	246.000	251.000	251.000	253.000	248.000	248.000	248.000	264.000	2
21. A urogenitalis DSM15434	241.000	241.000	242.000	245.000	264.000	252.000	255.000	264.000	253.000	257.000	266.000	256.000	254.000	2
22. A coleocanis DSM 15436	252.000	251.000	251.000	258.000	260.000	252.000	268.000	270.000	260.000	257.000	257.000	262.000	261.000	2

Other

In this case there are no identical sequences. Had there been, we would have deleted all but one of each set of identical sequences from the alignment in the Alignment Editor window, then saved the alignment again as `ebgC.mas` and exported it again as `ebgC.meg`.

Check Average Identity to Estimate Reliability of the Alignment

It should be noted that it is the multiple alignments, not the sequences themselves, that constitute the data from which trees are estimated. If the alignment is unreliable, so is the tree.

Codons: Pairwise amino acid identity

A study by Thompson et al. comparing a number of alignment programs showed that when the average percent amino acid identity in pairwise comparisons is too low, the accuracies of the multiple alignments fall below the level where they can produce reliable phylogenetic trees (Thompson et al. 1999).

Thompson's study showed that when the average amino acid identity is <20%, then <50% of the residues are correctly aligned. The 20% identity appears to be a significant threshold; in the "twilight zone" between 20% and 30% identity, about 80% of residues are correctly aligned, and above 30% identity more than 90% of residues are correctly aligned. Another study (Ogden and Rosenberg 2006) has shown that tree accuracy varies little with amino acid alignment accuracy as long as that accuracy is greater than 50%.

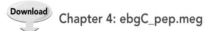 Chapter 4: ebgC_pep.meg

To determine the percent amino acid identity, click the **Translated Protein** tab and export that *protein* alignment as a new MEGA file, using the name `ebgC_pep.meg`.

In MEGA's main window first open the `ebgC_pep.meg` file, then choose **Compute Overall Mean Distance…** from the **Distances** menu. In the **Analysis Preferences** window be sure that **Substitutions Model** shows that the **Substitutions Type** is **Amino acid** and the **Model/Method** is **p-distance**, then click the **Compute** button (**Figure 4.11**).

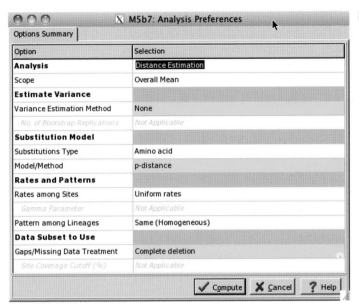

Figure 4.11

The p-distance is (1 – amino acid identity); thus if the average p-distance is <0.8 the alignment is acceptable; if ≥0.8 it is unreliable. In **Figure 4.12**, the average distance (highlighted) is 0.571, corresponding to 42.9% identity—well within the acceptable range.

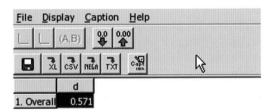

Figure 4.12

Non-coding DNA sequences

If your data are non-coding DNA sequences, the 20% amino acid identity criterion for sufficient accuracy does not apply. The corresponding figure for alignments of non-coding DNA is 66% sequence identity to ensure about 50% alignment accuracy (Kumar and Filipski 2006).

You measure DNA sequence identity the same way you measure amino acid identity: by determining the average distance under the p-distance model. If the average distance is more than 0.33, then identity is below 66% and accuracy is probably too low to permit using the alignment to estimate a valid phylogeny.

If the alignment is unreliable, do *not* go on to make a phylogenetic tree; the resulting tree would be meaningless to both you and your audience. To remedy the situation, remove the sequences that are most divergent from your sequence of interest until the amino acid p-distance is <0.8 (or the nucleotide p-distance is <0.33).

Chapter 12 will explore a more sophisticated way to evaluate the quality of alignments, but for the moment using the percent average identity will suffice. The important thing to keep in mind is that if the alignment is unreliable, the tree will be unreliable.

Increasing Alignment Speed by Adjusting MUSCLE's Parameter Settings

The temptation is to always accept the default values because we usually don't know what the various parameters signify or why those particular values have been assigned to them. Although default settings usually work well, it is useful to understand them and to understand a bit more clearly just how MUSCLE works. In particular, some parameters can be adjusted to significantly decrease the time required for MUSCLE to align large data sets.

How MUSCLE works

MUSCLE aligns sequences in a three-stage process:

- In the first stage, it determines the k-mer distances among all the sequences (a fast process because it does not require alignment), then uses a clustering method to construct a guide tree based on those distances. And finally, MUSCLE carries out a progressive multiple alignment of the sequences in an order dictated by the guide tree.

- In its second stage, MUSCLE improves the progressive alignment by repeating the first stage, using distances estimated from the first-stage multiple alignment to make the second-stage guide tree, then using that guide tree to make a new second-stage multiple alignment.

- The third stage further refines the second-stage alignment.

During the progressive alignments, MUSCLE introduces gaps into each sequence in an attempt to maximize the number of characters that match. As it does so it assigns a positive score for each match, and the score for the alignment is the sum of those individual character match scores. MUSCLE seeks the alignment that maximizes the score. If we could introduce as many gaps as we please, we could write any two completely unrelated sequences one above the other, in such a way that each character was either above an identical character or above a gap. The result would be a perfect—but meaningless—score. The solution to the problem is to reduce the score for each gap. Usually there is a large penalty for opening a gap, and a smaller penalty for each additional gap character within the gap. Thus gaps that reduce the score by more than is gained by additional matching characters are not introduced.

The various parameters that can be set control how MUSCLE executes each of the three stages.

Adjusting parameters to increase alignment speed

If you are interested in understanding MUSCLE's parameter settings you should read the article by Edgar (Edgar 2004a). As should be expected, there is a trade-off between speed and accuracy. Edgar discusses three settings: the default settings, which are optimized for accuracy; the "Muscle-fast" settings, which are optimized for speed; and the "Muscle-prog" settings, which are a compromise between accuracy and speed. The default settings are those shown in **Figure 4.13**.

Figure 4.13

Option	Selection
☐ Presets	None
Gap Penalties	
☐ Gap Opening	-12
☐ Gap Extension	-1
☐ Hydrophobicity Multiplier	1
Memory/Iterations	
☐ Max Memory in MB	1449
☐ Max Iterations	16
More Advanced Options	
☐ Clustering Method (Iteration 1,2)	UPGMB
☐ Clustering Method (Other Iterations)	UPGMB
☐ Max Diagonal Length	24
☐ Other Commands	
☐ Genetic Code (when using cDNA)	Standard
Alignment Info	MUSCLE stands for multiple sequence comparison by log-expectation. It is a public domain multiple alignm

To increase speed, check the **Presets** box, then click the triangle that appears at the right end of that line to display the choices shown in **Figure 4.14**. **Fast speed** is the equivalent of "Muscle-fast" and **Large Alignment** is the equivalent of "Muscle-prog."

Figure 4.14

Option	Selection
☑ Presets	None
Gap Penalties	None
☐ Gap Opening	Large Alignment (Max iterations = 2)
	Fast Speed (Max iterations = 1, diagonals, kbit20 3)
☐ Gap Extension	Refining Alignment
☐ Hydrophobicity Multiplier	1
Memory/Iterations	
☐ Max Memory in MB	1398
☐ Max Iterations	16
More Advanced Options	
☐ Clustering Method (Iteration 1,2)	UPGMB
☐ Clustering Method (Other Iterations)	UPGMB
☐ Max Diagonal Length	24
☐ Other Commands	
☐ Genetic Code (when using cDNA)	Standard
Alignment Info	MUSCLE stands for multiple sequence comparison by log-expectation. It is a public domain multiple alig

Depending on the data set, Muscle-fast is anywhere from 7.4 to 33 times faster than MUSCLE (4.6 to 27 times faster than ClustalW), but at a cost of a 5% reduction in accuracy relative to MUSCLE. Muscle-fast is as accurate as ClustalW, however.

Muscle-prog is only 1.5 to 3.6 times faster than MUSCLE (and 0.5 to 5 times faster than ClustalW), but at a sacrifice of only 1% in accuracy relative to

MUSCLE and remaining about 4% more accurate than ClustalW. Only you can decide when these trade-offs are worth it.

The default clustering method is UPGMB. You can change that to Neighbor Joining, and after reading *Learn More about Distance Methods* (p. 64) you may be tempted to do so. Resist that temptation while aligning sequences. Edgar (2004a) points out that while Neighbor Joining makes more accurate phylogenetic trees, it gives worse guide trees for alignment purposes.

Aligning Sequences with ClustalW

Aligning nucleotide or protein sequences with ClustalW is very similar to aligning them with MUSCLE. Coding sequences should be aligned choosing **Align by ClustalW** from the **Alignment** menu or by clicking the **Align selected blocks by ClustalW** icon (circled in **Figure 4.15**), then choosing **Align Codons**.

Figure 4.15

The major difference is the appearance of the Clustal Parameters window (**Figure 4.16**). Notice that when either aligning DNA coding sequences by codons or when aligning protein sequences, the window indicates **Protein** (circled). When aligning DNA sequences directly the default parameter setting are fine, but when aligning codons or protein sequences accuracy can be improved by changing the **Multiple Alignment** parameters (boxed in Figure 4.16) to **Gap Opening penalty = 3.0** and **Gap Extension Penalty = 1.8**, as shown.

I have emphasized—several times—that a phylogenetic tree is only as good as the quality of the alignments the tree is based on. Now that I have discussed how to achieve the most accurate alignments, the next several chapters deal with the various algorithms for creating trees from your (now finely honed) alignment data.

Figure 4.16

Protein

Pairwise Alignment

| Gap Opening Penalty | 10 |
| Gap Extension Penalty | 0.1 |

Multiple Alignment

| Gap Opening Penalty | 3 |
| Gap Extension Penalty | 1.8 |

Protein Weight Matrix	Gonnet
Residue-specific Penalties	ON
Hydrophilic Penalties	ON
Gap Separation Distance	4
End Gap Separation	OFF
Genetic Code Table:	Standard

| Use Negative Matrix | OFF |
| Delay Divergent Cutoff (%) | 30 |

☐ Keep Predefined Gaps

| Specify Guide Tree | | ... |

| ? Help | ✓ OK | ✗ Cancel |

CHAPTER 5

Major Methods for Estimating Phylogenetic Trees

You may already be aware that there are a number of methods currently being used to estimate trees from sequence data, and you may also be aware that the field of phylogenetics is quite contentious with respect to which method is best. If you ask an evolutionary colleague which method to use, you are likely to get an answer such as, "You *must* use Parsimony" (or Neighbor Joining, or Maximum Likelihood, etc., depending on which colleague you ask). Much of what you will hear is akin to religious conviction, and you need not worry about it. You could simply stick with the Neighbor Joining (NJ) method used in the Chapter 2 Tutorial. This is indeed a useful and widely used method, and I will discuss NJ in detail in Chapter 6; however, other methods offer some advantages (and some disadvantages) when compared with NJ. It is important to understand several methods and to make your choices based on the situation at hand rather than limiting yourself to NJ.

There are two primary approaches to tree estimation: **algorithmic** and **tree-searching**. The algorithmic approach uses an algorithm to estimate a tree from the data. The tree-searching method estimates many trees, then uses some criterion to decide which is the best tree or best set of trees (see *Learn More about Tree-Searching Methods*, p. 62).

The algorithmic approach has two advantages. It is fast, and it yields only a single tree from any given data set. Neighbor Joining is one of the two algorithmic methods in current use; the other is UPGMA (Unweighted Pair-Group Method with Arithmetic Mean). NJ has almost completely replaced UPGMA in the current literature.

All the other currently used approaches are tree-searching methods. They are generally slower, and some will produce several equally good trees. At first it may seem that the algorithmic methods are the obvious choice—they are fast and they result in a single tree that you can publish, so you can quickly get on with other things. At one time the speed issue was important, especially when a data set included many sequences. However, today's powerful desktop computers have greatly reduced the speed problem, so for most data sets the speed advantage of algorithmic methods is negligible.

LEARN MORE ABOUT

Tree-Searching Methods

Methods such as Parsimony, Maximum Likelihood, and Bayesian analysis search for the tree that best meets some optimality criterion by evaluating individual trees. When the number of taxa is small, it is possible to evaluate each of the possible trees—that is, to conduct an exhaustive search that guarantees finding the best tree because every tree has been evaluated. But with even 10 taxa, there are more than *34 million* rooted trees (see *Learn More about Phylogenetic Trees*, p. 72) and an exhaustive search is both impractical and, well, exhausting.

An exhaustive search is carried out by finding each of the possible trees by a branch-addition algorithm. The first three taxa are connected to form the only possible three-taxon tree, one that contains three branches (tree A in Figure 1). The fourth taxon is added by adding a new branch to the middle of each of the existing branches to generate the three possible four-taxon trees (trees B1, B2, and B3).

Adding the fifth taxon requires adding a new branch to the middle of each of the five branches in each of the four-taxon trees to generate 15 trees. This is accomplished by adding each of the five possible branches to tree B1 to estimate trees $C1_1$–$C1_5$, then backing down to tree B2 and adding each of the five branches to make trees $C2_1$–$C2_5$, then backing down to tree B3 and again adding the five possible branches to make trees $C3_1$–$C3_5$. If there were six taxa, starting with tree $C1_1$ and going through tree $C3_5$ seven branches would be added to each tree to make all of the possible trees at the D level.

An alternative, the **branch-and-bound** algorithm, also guarantees finding the best tree but does not require searching every tree. A random tree containing all taxa is generated and evaluated. Then, starting at A (the three-taxon tree in Figure 1), the search moves out toward the tips. It does not attempt to construct

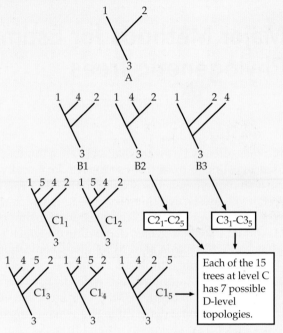

FIGURE 1

all possible trees at each level of the search; instead it constructs a single tree (say, B1) and evaluates it. If the criterion is minimum evolution and the current tree has a better (lower) score than the random starting tree, the search moves on to the next level by adding another branch. If the current tree has a worse score than the random tree, then it and all other trees that can be derived from it by adding more branches will have worse scores. The branch-and-bound search can thus discard all of its descendants without evaluating them. When that occurs, the search backs up one level, adds a branch somewhere else, and again starts searching toward the tip.

If the search gets all the way to the tip and finds a score that is better than that of the random tree, that score now becomes the score against which all other scores are judged.

As in the exhaustive search, the entire tree is covered by eventually backing down to the root level and starting out along the path that begins with B2, and then along the path that begins with tree B3. When the number of trees is large and evaluating each tree would be too slow to permit using the branch-and-bound algorithm, a **heuristic** strategy is used. A heuristic approach is essentially a hill-climbing algorithm in which an initial tree is selected, then rearrangements are sought that improve the tree.

There are too many heuristic algorithms to describe them in detail, but one common approach (with many variants) is the **stepwise addition** method. It is similar to branch-and-bound in that it starts with a three-taxon tree, then adds branches to make each of the three possible four-taxon trees. The difference is that at this point each of the trees is evaluated and the one with the best score is selected to make the five possible five-taxon trees that can be derived from it. At each level, only the best of the trees at that level is used to add the next taxon.

However, if tree B2 is the best tree at the B level, but the best tree *containing all of the taxa* happens to be derived from tree B3, then the best tree will not be found. In effect, the step-wise search will climb to the top of a hill, but it will not necessarily be the highest hill. Some algorithms try to avoid that by starting with the most divergent taxa, but there is no method that guarantees finding the best tree.

Often stepwise addition is used to find an initial tree to which **branch swapping** is then applied. Branch swapping involves making predefined rearrangements of trees by one of a variety of means.

An alternative to stepwise addition is **star decomposition**. Initially, all of the taxa are joined at a single internal node. Then a set of trees is generated by each of the possible joinings of a pair of taxa at a node leading away from that central node. Each of those trees is evaluated by the optimality criterion and the best is used for the next step.

Because they do not guarantee finding the best tree, heuristic searches always involve a trade-off between the certainty of finding the best tree and the need to find a tree within a realistic time.

Swofford et al. 1996, pp. 478–483, presents a detailed but quite readable discussion of methods for searching for optimal trees.

Although it may appear advantageous to have only a single tree to think about, that comfort can be misleading because it gives the false impression that the tree you see is the "right" one. It is essential to understand that the "right" tree doesn't exist. We are trying to deduce the order in which existing taxa (by which here we mean sequences) diverged from a hypothetical common ancestor, and to calculate the amount of change along the branches between the diverging events. It is extremely unlikely that those deductions will be correct in every detail, so the tree we see will not be an accurate depiction of historical events. Even if we are only concerned with tree topology, we can never be assured that the topology of the tree accurately reflects the historical branching order. Whatever method we choose, the only thing we can be sure of is that the resulting tree is wrong. The best we can hope for is the correct tree—i.e., the tree that *most closely approximates* what happened in the past.

In short, we must recognize that since we don't know what happened in the past, we can never be entirely sure how accurate the tree is. Tree-searching methods may yield one tree or several trees, but all methods implicitly acknowledge that the trees produced are only a subset of the possible trees that are consistent with the data.

Distance versus Character-Based Methods

The two algorithmic methods, **Neighbor Joining** and **UPGMA**, are both *distance methods*. Distance methods convert aligned sequences into a distance matrix of pairwise differences (distances) between the sequences (see *Learn More about Distance Methods*, below). The matrix is much like the tables of "percent homology" that often appear in the literature when only a few sequences are being compared. Distance methods use that matrix as the data from which branching order and branch lengths are computed.

Character-based methods include Parsimony, Maximum Likelihood, and Bayesian Inferrence. All use multiple alignment directly by comparing characters within each column (each site) in the alignment.

- **Parsimony** looks for the tree or trees with the minimum number of changes (*Learn More about Parsimony*, p. 111). It is often the case that there are several trees, differing only slightly, that are consistent with the same number of events, and that are therefore equally parsimonious.

LEARN MORE ABOUT

Distance Methods

In distance methods, distances are expressed as the fraction of sites that differ between two sequences in a multiple alignment. It is fairly obvious that a pair of sequences differing at only 10% of their sites are more closely related than a pair differing at 30% of their sites. It also makes sense that the more time has passed since two sequences diverged from a common ancestor, the more the sequences will differ. Although the latter assumption is reasonable, it is not always true. It might be untrue because one lineage evolved faster than the other. Even if two lineages evolved at the same rate, the assumption might be untrue because of multiple substitutions.

As two sequences diverge from a common ancestor, each nucleotide substitution initially will increase the number of differences between the two lineages. As those differences accumulate, however, it becomes increasingly likely that a substitution will occur at the same site where an earlier substitution occurred. For example, if a site that was an A in the ancestor does not change in one lineage but in the other lineage changed from A to C and then later, along the same branch, changed back to A, what were actually two changes will be counted as zero.

If in one lineage the site changed from A to C then later to G, two changes will be counted as one change. And if the same site changes from A to C in both descendant lineages, again two changes will be counted as zero. Because the number of observed differences almost always underestimates the actual amount of change along lineages, a variety of models are used to estimate corrected distances from the number of observed differences.

The two most popular distance methods, UPGMA and Neighbor Joining, are both algorithmic methods (i.e., they use a specific series of calculations to estimate a tree). The calculations involve manipulations of a distance matrix that is derived from a multiple alignment. Starting with the multiple alignment, both programs calculate for each pair of taxa the distance, or the fraction of differences, between the two sequences and write that distance to a matrix.

UPGMA

UPGMA (Unweighted Pair-Group Method with Arithmetic Mean) is an example of a clustering method. The program first finds the pair of taxa with the smallest distance between them and defines the branching between them as half of

- **Maximum Likelihood** looks for the tree that, under some model of evolution, maximizes the likelihood of observing the data (*Learn More about Maximum Likelihood*, p. 123). ML almost always recovers a single tree, but some programs can be instructed to save multiple trees. One advantage of the ML method is that the likelihood of the resulting tree is known.

- **Bayesian Inference** is a recent variant of Maximum Likelihood. Instead of seeking the tree that maximizes the likelihood of observing the data, it seeks those trees with the greatest likelihoods given the data (see *Learn More about Bayesian Inference*, p. 141). Instead of producing a single tree, Bayesian inference produces a set of trees with roughly equal likelihoods. The results of a Bayesian analysis are easy to interpret because the frequency of a given clade in any set of trees is virtually identical to the probability of that clade existing. Thus, no bootstrapping is necessary to assess the confidence in the structure of the tree.

that distance—in effect placing a node at the midpoint of the branch. It then combines the two taxa into a "cluster" and rewrites the matrix with the distance from the cluster to each of the remaining taxa. Since the "cluster" serves as a substitute for two taxa, the number of entries in the matrix is now reduced by one. The process is repeated on the new matrix and reiterated until the matrix consists of a single entry. That set of matrices is then used to build up the tree by starting at the root and moving out to the first two nodes represented by the last two clusters.

UPGMA has built into it an assumption that the tree is additive (see *Learn More about Phylogenetic Trees*, p. 70) and that it is ultrametric (i.e., that all taxa are equally distant from a root—an assumption that is likely to be incorrect). The same, usually false, assumption is used to place the root of a UPGMA tree. For that and other reasons, UPGMA is rarely used today by phylogeneticists, but its use in microbial epidemiology studies remains fairly common. I strongly discourage the use of UPGMA to estimate phylogenetic trees.

Neighbor Joining

Neighbor Joining (NJ) is similar to UPGMA in that it manipulates a distance matrix, reducing it in size at each step, then reconstructs the tree from that series of matrices. It differs in that NJ does not construct clusters but directly calculates distances to internal nodes.

From the original matrix, NJ first calculates for each taxon its net divergence from all other taxa as the sum of the individual distances from the taxon. It then uses that net divergence to calculate a corrected distance matrix. NJ then finds the pair of taxa with the lowest corrected distance and calculates the distance from each of those taxa to the node that joins them. (The distances from the two taxa to the node need not be identical.) A new matrix is then created in which the new node is substituted for those two taxa. NJ does not assume that all taxa are equidistant from a root.

NJ is, like parsimony, a minimum-change method, but it does not guarantee finding the tree with the smallest overall distance. Indeed, there are cases in which many shorter trees than the NJ tree exist (Hillis et al. 1996). Some authors think that the best use of an NJ tree is as a starting point for a model-based analysis such as Maximum Likelihood (Hillis et al. 1996; Swofford et al. 1996), but I do not think you should automatically discount publishing an NJ tree.

Swofford et al. 1996 presents a detailed discussion of distance methods, and Hillis et al. 1996 includes a good discussion of comparing various methods. Li 1997 has a good discussion of the multiple-hits problem on pp. 69–70.

Chapter 6 and Chapters 8–10 cover each of the above methods in detail. In each chapter I illustrate the method with the SmallData data set (34 sequences and 498 sites), which is the *ebgC* alignment that was generated in Chapters 3 and 4 plus an additional five *E. coli* sequences to better illustrate some points. I also provide a tree based on a large data set (LargeData) consisting of 77 sequences and 1464 sites.

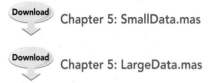

Download   Chapter 5: SmallData.mas

Download   Chapter 5: LargeData.mas

Which Method Should You Use?

As Nei and Kumar (2000) point out, choices among the various methods are often based on personal preference and scientific background. Because the field of systematics frequently uses morphological characters, scientists trained in that field often prefer parsimony and are suspicious of methods that employ mathematical models. Molecular biologists and geneticists prefer analytical methods, but do not trust sophisticated mathematical models. Those trained in mathematics and statistics tend to consider phylogenetics as a mathematical problem rather than a practical tool. We all bring biases to our choice of methods, but we would also like to judge those measures in some objective fashion.

It would be lovely if there were some objective way to select the "best" method for estimating evolutionary trees, but there isn't. No method is ideal for all performance criteria. Some of the criteria that have been considered are efficiency, robustness, computational speed, and discriminatory ability. *Efficiency* is a measure of how quickly the method converges on the correct tree as the amount of data (lengths of the sequences) increases; *robustness* is a measure of how well the method can tolerate deviations from its assumptions and still recover the correct tree; *computational speed* is just what the name says; and *discriminatory ability* is how well the method guarantees recovering the correct tree. There are trade-offs among these criteria due to the fact that a method that increases accuracy according to one measure may decrease accuracy according to another measure (Hillis et al. 1996).

You will have to decide, in each case, which method best suits your needs. I say "in each case" because you probably will not always choose the same method. If you were compulsive, you might use all four methods. I suggest that you consider three factors when picking a method: accuracy, ease of interpretation, and time.

Accuracy

One definition of accuracy is the probability that the method recovers the true tree. One might ask, "If we don't know which tree is the true tree, how can we measure how well a method recovers that tree?" Usually, with real data, we cannot. The exception is some experimental evolutionary systems in which all of the descendants of a single clonal organism are available and the true tree can be known.

Attempts to measure the relative effectiveness of methods are usually based on simulations in which a computer generates descendants of some starting sequence according to some evolutionary model (see *Learn More about Evolutionary Models*, p. 75). In the end, not only is a set of tip sequences generated, but all of the intermediate steps are known—so the true tree *is* in fact known. Various methods are then compared to see which method best recovers this true tree, and under what conditions they do so. Most of the simulations used in such studies are pretty unrealistic; for instance, they usually do not include insertion or deletion mutations, thus avoiding the necessity of aligning the resulting sequences. Because poor alignment quality is a real source of inaccuracy in estimated trees, the results of such simulations are difficult to apply when choosing methods to estimate trees from real data.

Two studies (Hall 2005; Ogden and Rosenberg 2006) have used more biologically realistic simulations to compare the various methods. Instead of asking the probability that a method recovers the true tree, those studies ask, "How close is the estimated tree to the true tree?" Both topological accuracy (i.e., the fraction of true clades that are present on the estimated tree) and branch length accuracy are considered. Despite somewhat different approaches to the problem, the two studies agree that Bayesian Inference is slightly more accurate than Maximum Likelihood, that Maximum Parsimony is next, and that Neighbor Joining is the least accurate approach.

Ease of interpretation

Neighbor Joining creates a single, strictly bifurcating tree that fails to convey uncertainty of branching order very well, even when considerable uncertainty exists. Maximum Parsimony often yields multiple trees. A consensus tree can convey uncertainty in branching order as polytomies (see *Learn More about Phylogenetic Trees*, p. 70), but you can't put branch lengths onto that consensus tree. Maximum Likelihood returns just one tree, the most likely tree. Bayesian Inference gives you a consensus tree in which polytomies indicate uncertainty about branching order, but also gives you branch lengths. You need to consider which method you and your audience will be most comfortable with.

Time and convenience

Choosing among the methods is often just a pragmatic matter: If your computer takes more time than you are willing to spend to calculate the tree, then use a faster method. My own rule of thumb is that I am willing to use a method that will run overnight. Therefore, if it takes longer than about 14 hours, I will probably choose another method. Some phylogeneticists, however, have taken me to task for being obsessed with speed.

The extent to which speed matters depends both on your patience and your point of view. If the phylogenetic analysis is the main point of your paper, speed will probably not be a major consideration. Indeed, I have published

TABLE 5.1 Comparison of times required for the four major phylogenetic methods[a]

DATA SET	NEIGHBOR JOINING	PARSIMONY	MAXIMUM LIKELIHOOD	BAYESIAN INFERENCE
SmallData (without reliability estimate)	4 sec	26 sec	35 sec	—
SmallData (with reliability estimate)	56 sec	1 min 25 sec	59 min	2 hr 44 min
LargeData (without reliability estimate)	4 sec	1 min 22 sec	2 min 49 sec	—
LargeData (with reliability estimate)	1 min 43 sec	11 min 38 sec	3 hr 42 min	14 hr 29 min

[a]The analyses were carried out on a Macintosh MacPro Dual Processor computer running two Intel Xenon dual-core processors at 2.6 GHz and the OS10.6 operating system. Other computers will yield faster or slower times. Neighbor Joining, Parsimony, and Maximum Likelihood trees were estimated by MEGA 5.0. Neighbor Joining trees were estimated using the Maximum Composite Likelihood model, and tree reliability was estimated from 1000 bootstrap replicates. Parsimony trees (topology only) were estimated using the CNI-level 3 algorithm and tree reliability was estimated from 100 bootstrap replicates. Maximum Likelihood trees were estimated under GTR+G for the SmallData and GTR+G+I for the LargeData, and the bootstrap method with 100 replicates was used to estimate tree reliability. MEGA does not report the time required to estimate trees, but times were estimated manually as the time required for the Tree Explorer window to appear after clicking the Compute button. Bayesian Inference was by MrBayes 3.2 under the same models as for ML, for a sufficient number of generations for the standard deviation of split frequencies to fall below 0.01, and using the ML tree as a "user tree."

trees that took 2 weeks to estimate. On the other hand, if the analysis is just one minor aspect integrated into a largely experimental paper, speed may be very important to you.

Because for better or worse time is often the basis for deciding which method to use, I have applied all four methods to the same two data sets, with the results seen in **Table 5.1**. Of course, it is not only the time required to estimate the tree that matters; it is also the time required to estimate the *reliability* of the tree. I have therefore included times required to estimate the tree with and without estimating the reliability of the tree. The data set SmallData (*ebgC*) consists of 34 sequences with 498 sites. The LargeData set consists of 77 sequences with 1464 sites. Keep in mind that the times shown are relative; actual times will depend on the computer you are using and on other tasks that might be running in the background.

In the end, it may matter little which method you use. Neighbor Joining, Parsimony, Maximum Likelihood, and Bayesian Inference are all perfectly respectable methods that will be accepted by most journals and most readers as valid. If your data are good (in the sense that the sequences you chose really are related by descent) and your alignment is robust, then the trees will be so similar that the differences won't matter. In effect, the differences represent real uncertainty. If the differences are essential to your interpretation of the data, then you need to point out the uncertainty about those interpretations.

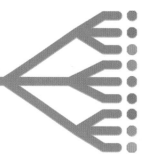

CHAPTER **6**

Neighbor Joining Trees

Neighbor Joining (NJ) is the most widely used of the distance algorithms (see *Learn More about Distance Methods*, p. 64). Its popularity is probably accounted for both by its speed and its simplicity. NJ produces a single, strictly bifurcating tree (see *Learn More about Phylogenetic Trees*, p. 70). *Strictly bifurcating* means that each internal node has exactly two branches descending from it.

Using MEGA 5 to Estimate a Neighbor Joining Tree

In my judgment, MEGA is the best program that is currently available for estimating NJ trees. The first step is to open a data file (an alignment) that is in the .meg format. In this case, open the `SmallData.meg` file from the MEGA main window as described in Chapter 4 (see Figure 4.8A).

Download

Chapter 6: SmallData.meg

Two buttons are now displayed in MEGA's main window. One is labeled **TA**, the other **Close Data** (**Figure 6.1**). The **TA** button shows that there is an active data file, and the name of the active file (circled) is always shown at the bottom right of the window. The **Close Data** button will close that file and cause both buttons to disappear.

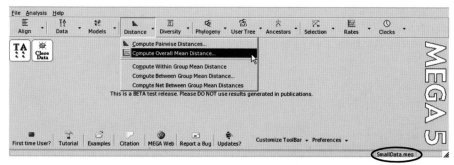

Figure 6.1

Phylogenetic Trees

A tree is a simple structure composed of *nodes* and *branches*. There are two types of nodes. **External nodes** are located at the tips of the tree and represent the taxa (or, in the case of sequence data, the molecular sequences) that exist today and that we can actually examine. **Internal nodes** represent ancestral taxa, whose properties we can only infer from the existing taxa. Nodes are connected by **branches**, lines whose lengths represent the amount of genetic change that occurred between an ancestral node and its descendant.

As illustrated in Figure 1, numbers on the branches represent their respective **branch lengths**, usually expressed as **changes per site** in the aligned sequences. For example, between nodes X and Y in Figure 1, 0.03 changes per site occurred, whereas only 0.01 change per site occurred between nodes Y and D. Even if exact values are not provided, the relative lengths of the branches may be drawn in proportion to the number of changes along that branch.

The tree in Figure 1 is **additive** because the distance between any two nodes equals the sum of the lengths of all the branches between them. While it might seem intuitive that all trees must be additive, it is not the case. If multiple substitutions have occurred at a particular nucleotide site, then additivity will not hold unless the distances are corrected for multiple substitutions.

A node is **bifurcating** if it has only two immediate descendant lineages. We usually assume that evolutionary speciation is a binary process resulting in the formation of two species from a single ancestral species. That may not always be the case, or available data may not make it possible to resolve the order in which species descended from a single common ancestor, in which case a node is **multifurcating**. A multifurcating node is more commonly called a **polytomy**.

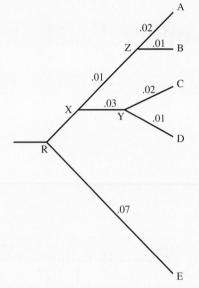

FIGURE 1 A rooted tree whose tips represent five taxa (A–E) in a clade. Four internal nodes (R, X, Y, Z) represent ancestral taxa, including the root (R). The numbers on the branches indicate the amount of change in a particular sequence that occurred along that branch.

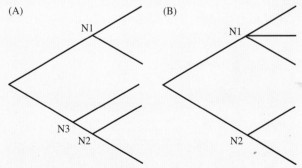

FIGURE 2 (A) A strictly bifurcating tree. Because nodes N1, N2, and N3 are all bifurcating, the tree is strictly bifurcating. (B) Because there is a polytomy at node N1 (i.e., node N1 has three descendant lineages), the tree is not strictly bifurcating.

Rooted and Unrooted Trees

A tree is said to be *rooted* if there is a particular node—the **root**—from which a unique directional path leads to each extant taxon. In Figure 1, R is the root because it is the only internal node from which all other nodes can be reached by moving forward (i.e., toward the tips). The root is the common ancestor of all of the taxa in the analysis.

An *unrooted* tree specifies only the relationships among the taxa; it has no directionality and does not define any evolutionary pathway. For any four taxa there are only the three possible unrooted trees seen in Figure 3A. Once a root is identified, five different rooted trees can be created for *each* of these unrooted trees (Figure 3.B). There are thus 15 possible

rooted trees for four taxa, each with a distinctive branching pattern reflecting a different evolutionary history for the relationships in the unrooted tree.

The number of possible trees, both rooted and unrooted, increases dramatically as the number of taxa increases. Where *s* is the number of taxa, the number of possible unrooted trees is

$$\frac{(2s-5)!}{2^{s-3}(s-3)!}$$

and the number of possible rooted trees is

$$\frac{(2s-3)!}{2^{s-2}(s-2)!}$$

Shown in tabular form (see next page), the results are startling.

As databases have grown, trees of >100 sequences—more than 100 taxa for molecular phylogenetic purpose—have become increasingly common.

Most phylogenetic methods produce unrooted trees, but unless you specifically choose MEGA's "Radiation" (unrooted) format

FIGURE 3 (A) For any four taxa, there are three possible unrooted trees. For each of the three unrooted trees, five distinct rooted trees can be derived by the choice of where to place the root. (B) The five trees derived from the leftmost tree in (A), rooted, respectively, in the branch between the A–B and C–D clades; at taxon A; at taxon B; at taxon C; and at taxon D.

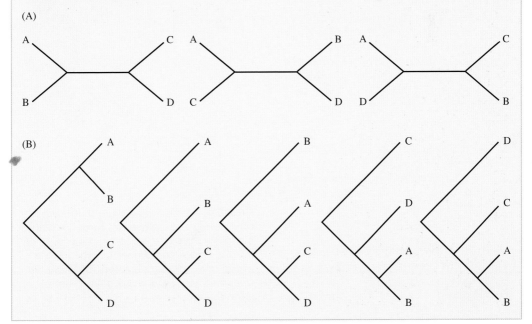

TAXA	UNROOTED TREES	ROOTED TREES	COMMENT
4	3	15	
8	10,395	135,135	
10	2,027,025	34,459,425	
22	3×10^{23}		Almost a mole of trees!
50	3×10^{74}		About the number of atoms in the universe
100	2×10^{182}		

to display the tree, when MEGA prints those trees they appear to be rooted. For display purposes, MEGA has put a bend in one branch or another by the midpoint rooting method, but that does not accurately root an unrooted tree. (See Chapter 7 for a more detailed discussion of rooting trees.)

Unless you know that a tree is properly rooted, either because you rooted it yourself or because an author specifically tells you it is rooted, assume that it is unrooted. Printing a tree in an unrooted or radiation format makes the tree's unrooted status absolutely clear.

Distinguishing Information from Appearance

It is important to distinguish between the relationships a tree depicts and the way that tree is drawn. For instance, it is obvious that the two trees in Figure 4A are the same and that no phylogenetic information changes as the result of swapping the way the branches to taxa A and B were drawn. It is not as obvious, but it is also true that the two trees in Figure 4B depict the same phylogeny.

The trees in Figure 4 are the result of placing a root into different branches of the first (top) unrooted tree in Figure 3A. It is less obvious that the two trees in Figure 4B are the same rooted tree. However, in both trees one branch leads from the root to taxon A and the other branch leads from the root to the BCD clade. Likewise, in both trees the BCD clade has one branch that leads to B and the other branch that leads to the CD clade. The fact that there are many ways to draw exactly the same tree often makes it difficult to judge the similarity of trees by eye.

A succinct overview of phylogenetics emphasizing the molecular aspects can be found in Chapter 5 of Graur and Li 2000.

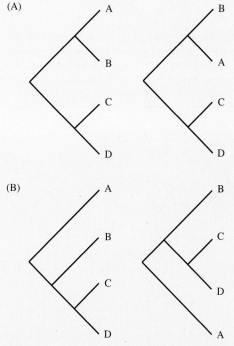

FIGURE 4 (A) Despite a slight difference in their appearance, these two trees are quite obviously the same; switching the positions of taxa A and B does not change the evolutionary history depicted. (B) These two trees also convey identical phylogenetic information.

Click the **TA** button to display the **Sequence Data Explorer** window (**Figure 6.2**). The top line (red arrow) shows the bases of the first sequence listed, in this case **E coli K12 MG1655**. The remaining individual sequences only show a base if it is different from that of the first sequence; otherwise they show a dot. To see all of the bases, click the **TA** button in the **Sequence Data Explorer** toolbar. Doing so is helpful when you want to get an impression of the number of gaps in the alignment. **It is essential to visualize all the bases if you want to export the alignment in another format.**

Figure 6.2

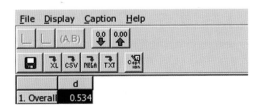

	A	T	G	A	G	G	A	T	C	A	T	C	G	A	T	A	A	C	T	T	A	G	A	A	C	A	G	T	T	C	C	G	C	C	A
☑ 1. E coli K12 MG1655																																			
☑ 2. S flexneri 2a str301																																			
☑ 3. E coli O157H7 EDL933																																			
☑ 4. E doacae subsp. doacae NCTC9394				T	C	G		T	C		T			G		G		C		G									T		A	A			
☑ 5. P damselae subsp. damselae				T	C	G		A	C		T										G			A			T	A	A	A	T	C			
☑ 6. V harveyi 1DA3				T	C	G		G	T		A			C										A			A	A	A	G	T				
☑ 7. V shilonii AK1				T	C				T		A			G		G								A			T	A	A	G	T	C			
☑ 8. V vulnificus YJ016				T	C	G		G	T		A			C		G					G			A			A	A	A						
☑ 9. V corallilyticus ATCC BAA450				T	C	G		G	T		A			A						C	G			A			A	A	A	G	T				
☑ 10. V parahaemolyticus AQ4037				T	C	G		G	T		A			C		G					G			A			A	A	A						
☑ 11. V splendidus 12B01				T	C	G		T	T		A			C				C			G			A			T	A	A	G	T				
☑ 12. V orientalis CIP102891				T	C	G		G	T		A			G		G					G			A			A	A	A	G	T				
☑ 13. P profundum SS9			G	T	T	G		T	T		A			G				C		C		G		A			A	A	A	A	C				
☑ 14. G hollisae CIP101886				T	C	G		G	T		A			C		G				G		C	G	A			A	A	A						
☑ 15. P angustum S14				T	C	G		A	T		A					G						C	G	A			T	A	A	T	T				
☑ 16. V bacterium SWAT-3				T	C	G		T	T		A			G										A			T	A	A	G	T				
☑ 17. V fischeri MJ11				T	C			T	T		A			C			T	C						A			T	A	A	G	T				
☑ 18. A hydrophila subsp hydrophila ATCC79				T	C		C		C		G			C					C		G	A	C	C		T			A	A	A	A			
☑ 19. S proteamaculans 568				T	T				T	G					A	T	C	A					C	G	G				T		A	G		T	

1/501	Highlighted: None	Data

Determine the suitability of the data for a Neighbor Joining tree

First you must determine whether the data are even suitable for estimating Neighbor Joining trees. In their book *Molecular Evolution and Phylogenetics*, Nei and Kumar (2000), the authors of MEGA, say that if the average pairwise Jukes-Cantor (JC) distance is >1.0, the data are not suitable for making NJ trees and another phylogenetic method should be used. To determine the average JC distance, find the **Distances** menu of the MEGA main window and choose **Compute Overall Mean** (see Figure 6.1).

In the resulting **Analysis Preferences Window**, set **Model/Method** to **Jukes-Cantor** and click the **Compute** button. For the `SmallData` file the average distance is 0.534 (**Figure 6.3**), which is quite suitable for making NJ trees.

Figure 6.3

File   Display   Caption   Help

	d
1. Overall	0.534

Estimate the tree

In the main MEGA window choose **Construct/Test Neighbor Joining Tree** from the **Phylogeny** menu.

The Analysis Preferences window (**Figure 6.4**) is used to set up the conditions for estimating the tree. The window shows that the selected analysis is Phylogeny Reconstruction, and that Neighbor Joining is the chosen method of reconstruction. The yellow areas allow you to set the options that determine the conditions for estimating the tree.

Figure 6.4

Options Summary	
Option	**Selection**
Analysis	Phylogeny Reconstruction
Scope	All Selected Taxa
Statistical Method	Neighbor-Joining
Phylogeny Test	
Test of Phylogeny	None
No. of Bootstrap Replications	*Not Applicable*
Substitution Model	
Substitutions Type	Nucleotide
Genetic Code Table	*Not Applicable*
Model/Method	Maximum Composite Likelihood
Fixed Transition/Transversion Ratio	*Not Applicable*
Substitutions to Include	d: Transitions + Transversions
Rates and Patterns	
Rates among Sites	Uniform rates
Gamma Parameter	*Not Applicable*
Pattern among Lineages	Same (Homogeneous)
Data Subset to Use	
Gaps/Missing Data Treatment	Complete deletion
Site Coverage Cutoff (%)	*Not Applicable*
Select Codon Positions	☑ 1st ☑ 2nd ☑ 3rd ☑ Noncoding Sites

✓ Compute   ✗ Cancel   ? Help

The first choice, **Test of Phylogeny**, will be ignored for the moment, but I will return to it later in the chapter.

The next choice, **Substitution Type**, has automatically been set to **Nucleotide**.

The third choice is **Model/Method**, and for the nucleotide mode you are offered no fewer than eight choices. The default choice is the **Maximum Composite Likelihood** model, and Sudhir Kumar, one of MEGA's authors, recommends that this model always be used. The Jukes-Cantor model corrects for multiple substitutions at the same site (see *Learn More about Evolutionary Models*, p. 75), the Kimura 2-Parameter model allows transition and transversion substitutions to occur at different rates, and the Tamura-Nei model corrects for base compositional bias that differs from the default frequency of 0.25. **Maximum Composite Likelihood** is a likelihood-based implementation of the Tamura-Nei model that increases the accuracy of calculating the pairwise distances. Like Kumar, I suggest sticking with Maximum Composite Likelihood.

Evolutionary Models

Models determine the way in which a program calculates branch lengths. **Branch lengths** are intended to indicate the amount of genetic change between an ancestor and its descendant, so it might seem that the obvious way to calculate branch lengths is based on the number of differences in the sequences. However, to permit comparisons between trees of the same taxa, but using different genes, we usually express branch lengths in terms of the *proportion* rather than the number of sites that have changed. That choice—using the observed proportion of differences between sequences—is the **p-distance model**. All of the other models use a variety of assumptions to estimate how many additional, unobserved, differences should be added in order to calculate branch lengths. P-distance usually makes the best sense to molecular biologists, who intuitively trust observations and don't like to muck about with them. That intuition, however, is not a good guide for the analysis of evolutionary history.

Sequences diverge from a common ancestor because mutations occur and some fraction of those mutations become fixed in the evolving population (whether by selection or by chance), resulting in the substitutions of one nucleotide for another at different sites. In order to reconstruct evolutionary trees, we must make some assumptions about the substitution process and state those assumptions in the form of a model. When you use MEGA to create an NJ tree or an ML tree, you must choose a model by name or accept the default model. When you use MrBayes to create a BI tree, you must set the values of several parameters to define the model. The purpose of models is to define the assumptions that are used to estimate branch lengths, but those estimates can also affect the topology of the tree.

Consider two 1000-bp sequences that have diverged from a common ancestor. Aligning those sequences, we observe differences at 500 sites, so the p-distance would be 0.5. However, if those sites changed randomly, some sites will have changed twice—say from A to G, then from G to T—even though we can observe only one difference. Similarly, if a site changes from A to G and then changes back to A, two changes have occurred but we will observe zero changes. Clearly, there is a potential for underestimating the amount of change that has occurred along a branch because of multiple substitutions at the same site.

Each of the models represents an attempt to account for multiple hits and various other aspects of the ways in which we think nucleotide substitutions occur during gene evolution.

One-Parameter Models

The easiest model to consider is one in which the probabilities of any nucleotide changing to any other nucleotide are equal. In order to predict the probability that a particular nucleotide at a particular site will change to some other specific nucleotide over some time interval, we need only know the instantaneous rate of change (i.e., the rate at which nucleotide substitutions occur). This simple model has only one parameter, the substitution rate, and is known as the **one-parameter model** or the **Jukes-Cantor model** (Jukes and Cantor 1969).

Figure 1 shows that when the observed distance is 0.49 changes per site, as the result of multiple changes at the same site, the actual distance is 0.7945 changes per site. Notice that the curve flattens out as it approaches 0.75 observed changes per site. This number represents the upper limit of changes that we can observe, because there are only four states of each base (A, C, G, or T). Once this level of observed change is reached, any further change will be invisible.

If we are only interested in relative branch lengths, does it really matter if we apply the Jukes-Cantor correction? The answer is "yes,"

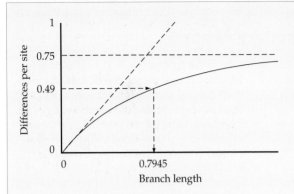

FIGURE 1 The expected difference per site between two sequences in the Jukes-Cantor model, as a function of branch length (the product of rate of change and time). (From Felsenstein 2004.)

for two reasons. First, as the upper limit of 0.75 in observed p-distance is approached, real differences in branch lengths will disappear, distorting the relative branch lengths. Second, underestimation of branch lengths can change the topology (or branching order) of a tree. In Figure 2 the unrooted tree on the left shows the true branching order and branch lengths. A and B share a most recent common ancestor, as do C and D. If those are the true branch lengths, the distances that would be observed are shown in the table in the center. The tree on the right is the one that would be estimated from those observed distances. Notice that not only are the distances between the taxa much less than on the true tree, the topology is also wrong. In the estimated tree the closest relative of B is C, not A.

The **Felsenstein 81** (**F81**) model extends the Jukes-Cantor model by allowing for differences in base frequencies. If we know that there is a G at some site at time $t = 0$, we can ask what the probability is that there will still be a G at that site at some later time t, and what the probability is that there will be, for instance, an A at that site instead. These are expressed, respectively,

as $P_{(GG)}(t)$ and $P_{(GA)}(t)$. If the substitution rate is a per time unit, then

$$P_{(GG)}(t) = \frac{1}{4} + \frac{3}{4}e^{-4at} \text{ and } P_{(GA)}(t) = \frac{1}{4} - \frac{1}{4}e^{-4at}$$

Since according to the one-parameter model all substitutions are equally likely, a more general statement is that

$$P_{(ii)}(t) = \frac{1}{4} + \frac{3}{4}e^{-4at} \text{ and } P_{(ij)}(t) = \frac{1}{4} - \frac{1}{4}e^{-4at}$$

When t is very close to zero, the probability that the site has not changed, $P_{(ii)}$, is very close to 1, while $P_{(ij)}$—the probability that the nucleotide at that site has changed from i to some other nucleotide, j—is close to 0. As time goes on, both probabilities approach 0.25; the time required for that approach depends on a.

FIGURE 2 An example of distortion of tree topology when uncorrected distances are used. The true tree is on the left; the uncorrected sequence differences under a Jukes-Cantor model are shown in the table in the center. The tree estimated from those differences, shown on the right, incorrectly separates B and C from A and D. (From Felsenstein 2004.)

	A	B	C	D
A	0.0	0.57698	0.59858	0.70439
B	0.57698	0.0	0.24726	0.59858
C	0.59858	0.24726	0.0	0.57698
D	0.70439	0.59858	0.57698	0.0

We can write a table that shows the instantaneous rates for each of the possibilities for change at a site:

Substituted Base

		A	C	G	T
	A	$-3a$	a	a	a
Original	**C**	a	$-3a$	a	a
base	**G**	a	a	$-3a$	a
	T	a	a	a	$-3a$

That table is usually expressed in the form of a matrix, commonly called the **Q matrix**, in which rows represent the original nucleotides and columns represent the nucleotide being substituted.

$$\mathbf{Q} = \begin{matrix} -3a & a & a & a \\ a & -3a & a & a \\ a & a & -3a & a \\ a & a & a & -3a \end{matrix}$$

Note that this is not a matrix of probabilities but a matrix of rates, and that the elements in a row sum to zero.

Other Models

The one-parameter model is the simplest, but it is not very realistic. We know that all changes do not occur at the same rates, and a variety of models have been proposed that allow us to specify different rates. The most general model is one in which each different substitution can occur at a different rate, involving 12 free parameters. Here the Q-matrix becomes:

$$\mathbf{Q} = \begin{matrix} -a -b -c & a & b & c \\ d & -d -e -f & e & f \\ g & h & -g -h -i & i \\ j & k & l & -j -k -l \end{matrix}$$

Other important models are all special cases of the above general model.

The **Kimura 2-parameter model** (Kimura 1980) extends the Jukes-Cantor correction by taking into account the possibility that the rates at which transitions and transversions occur might well be different. If all changes were equally probable, we should see twice as many transversions as transitions because

there are two possible transversions but only one possible transition. In reality, we often observe that transitions exceed transversions. This is probably because many transitions are silent, while most transversions involve amino acid substitutions that are likely to be selected against. So the **Kimura 2-parameter (K2P)** model calculates branch lengths based on two different rates. In Kimura's 2-parameter model, transitions occur at one rate, a, and transversions occur at a different rate, b:

$$\mathbf{Q} = \begin{matrix} -a -2b & b & a & b \\ b & -a -2b & b & a \\ a & b & -a -2b & b \\ b & a & b & -a -2b \end{matrix}$$

The **Tamura 3-parameter** model adds a correction for compositional bias. If the base ratios differ greatly from equal frequencies, perhaps because of mutational bias, then that difference needs to be accounted for. The **Tamura-Nei** model extends this by distinguishing between transitional substitution rates among purines and transversional substitution rates among pyrimidines. Other models that fall into this general family are the **Felsenstein 84 (F84)** model and the **HKY** model.

The **general time-reversible (GTR)** model (Tavaré 1986) has the rate matrix seen at the top of p. 78. There are six different substitution probabilities (e.g., $a = P[A \text{ to } C] = P[C \text{ to } A]$). It involves nine free parameters: $a, b, c, d, e, f,$ and the equilibrium base frequencies $\pi_A, \pi_C, \pi_G, \pi_T$ (which sum to one). Time-reversible models assume that $\pi_i q_{ij} = \pi_j q_{ji}$ for any i and j; that is, the amount of change from any bases i to j is the same in either direction. The substitution process looks the same whether we observe the process with time running forward or backward. An important consequence of the assumption is that one cannot identify the root of the tree without the assumption of the molecular clock.

When these evolutionary models are used to reconstruct trees, one may either assign specific values to those rates, or estimate the values from the data.

FIGURE 3   Rate matrix for the GTR model.

$$Q = \begin{bmatrix} -a\pi_C - b\pi_G - c\pi_T & a\pi_C & b\pi_G & c\pi_T \\ a\pi_A & -a\pi_A - d\pi_G - e\pi_T & d\pi_G & e\pi_T \\ b\pi_A & d\pi_C & -b\pi_A - d\pi_C - f\pi_T & f\pi_T \\ c\pi_A & e\pi_C & f\pi_G & -c\pi_A - e\pi_C - f\pi_G \end{bmatrix}$$

It is not uncommon in the phylogenetics literature to see a model specified as something like "K2P + InvGamma." That is simply jargon for "Kimura 2-parameter model with invariant sites estimated and with rate variation across sites estimated from a gamma distribution."

Rate Variation among Sites

The above models implicitly assume that the rates are the same at all sites, but it is also possible to include rate variation across sites in the models. It is easy to think of some examples in which evolutionary rates might be different at certain sites. For instance, the first codon in a gene encodes methionine, which almost always is ATG but, sometimes is GTG. We therefore expect that at sites 2 and 3 the substitution rate will be zero because substitutions at those sites will be strongly selected against. We also know that second positions in codons evolve most slowly, first positions evolve at an intermediate rate, and third positions evolve most rapidly; again because of selection. In general, positions on the interior of proteins—i.e., near active sites—evolve more slowly than do positions near the surfaces of proteins.

Because proteins fold, those slowly evolving sites tend to be located in patches rather than in one specific region of the sequence. These are examples of **site-specific variation**. One might imagine estimating individual rates for each site, but for typical genes that is not com-putationally practical when many sequences are involved. Also, as Felsenstein (2004) points out, with that many parameters, Maximum Likelihood often misbehaves. The alternative is to assume a well-behaved distribution of rates across sites; the distribution that is commonly used is the **gamma distribution**. It's not that the gamma distribution is particularly biologically realistic; rather, it is mathematically tractable (Felsenstein 2004).

Invariant Sites

As mentioned above, some sites, such as initiation codons, may not be free to vary at all. When protein function absolutely requires a tryptophan at a certain position, the sites in that codon would likewise not be free to vary. This constitutes a special case of site-specific variation. Often it is possible to examine an alignment and notice sites that are identical in all sequences. That does not necessarily mean that identical sites are invariant sites; it may be that the sequences are too closely related for any substitutions to have occurred at those sites. Both MEGA and MrBayes offer the option of estimating the proportion of invariant sites as part of the model specification.

Chapter 3 in *Fundamentals of Molecular Evolution* (Graur and Li 2000) offers a readable discussion of evolutionary models, while *Inferring Phylogenies* (Felsenstein 2004) offers a more detailed discussion of the topic.

Substitutions to Include allows you tease apart which kinds of substitutions are used to calculate the distances. The default is to use **Transitions+Transversion**, but you can also choose transitions only or transversions only. Unless you have a good reason to change this, stick with the default.

Rates among sites allows you to handle rate variation among sites (see *Learn More about Evolutionary Models*, p. 78). The default rate, **Uniform rates**, is usually sufficient. The alternative, **Gamma Distributed (G)**, requires that we specify a shape parameter. Again, the default value of 1 is usually sufficient, but MEGA does offer the possibility of calculating that rate from the data by clicking the **Rates** button and choosing **Estimate Gamma Parameter for Site Rates (ML)....** (**Figure 6.5**).

Figure 6.5

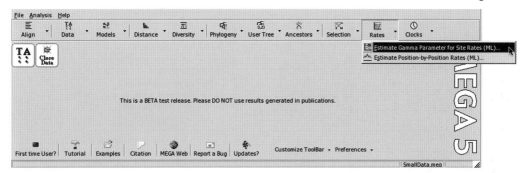

When an **Analysis Preferences** window appears just click the **Compute** button. A **Progress** window will open, and shortly a **Caption Viewer** window will appear within which the value for the shape parameter will be reported. If the reported value is much different from the default value you can enter the estimated value instead in the NJ **Analysis Preferences** window.

Returning to the choices in Figure 6.4, modifying **Patterns among Lineages** is beyond the scope of this book. Accept the default **Same (homogeneous)** setting.

Gaps/Missing Data, determines how NJ handles gaps. The default choice is **Complete Deletion**, which means that the program ignores all sites (columns in the alignment) that include a gap anywhere in the column. **Complete Deletion** is fine for this particular data set, which includes very few sites with gaps. However, for a data set in which there are many gaps, it is not a good choice because it removes a large fraction of the sites from consideration. If there is any doubt, I'd err on the side of caution and change this parameter to **Pairwise Deletion** so that sites with missing data are removed only as the need arises, instead of always being removed.

The last choice is **Select Codon Positions**. The default is to use all three positions, which is what you will usually do. It is possible to make a tree based only on the bases at the third position of each codon. Because of redundancy in the genetic code, many substitutions at third positions will be silent and therefore not subject to selection. If you want the best estimate of the relative rates of change along the branches, you may want to use only those third-position sites. However, I recommend that you use the default choice of all three codon positions unless you have a good reason to change it. The settings I suggest for the SmallData file were shown in Figure 6.4.

Once the settings are chosen, click the **Compute** button to estimate the tree. The resulting NJ tree will be displayed in the **Tree Explorer** window (**Figure**

6.6). If the entire tree is not visible resize the window if necessary, then click the **Fit Tree to Screen** button (red arrow in Figure 6.6).

Figure 6.6

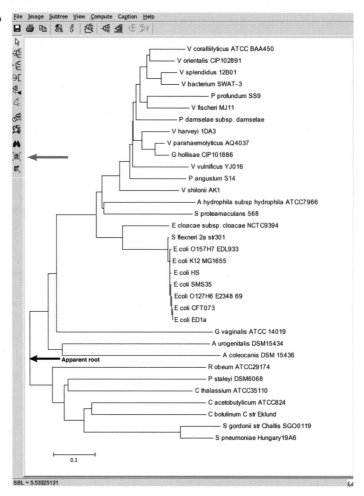

Unrooted and Rooted trees

The SmallData NJ tree shown in Figure 6.6 makes it appear that all of the sequences are descended from a common ancestor, which is represented by the leftmost interior node, indicated by the black arrow. That appearance is quite misleading. The interior node from which all the sequences or taxa are descended is called the **root** (see *Learn More about Phylogenetic Trees*, p. 71). If we know the root of a tree, we know the direction of evolution—that is, the order of descent of the sequences. We can trace a unique pathway from the root to any given tip sequence. All of the descendants of any particular interior node constitute a clade. Understanding the order of descent depends upon knowing where the root of a tree lies.

The SmallData NJ tree shown in Figure 6.6 *appears to be* rooted; all of the sequences seemingly are descended from a common ancestor, which is represented by the leftmost interior node (black arrow in Figure 6.6). That appearance is completely misleading, however. The Neighbor Joining method—in common with all of the methods for reconstructing phylogenetic trees that are discussed in this book—is unable to determine where the root lies. Thus, these methods reconstruct **unrooted** trees.

When a tree is drawn in the rectangular format, some node must be drawn at the extreme left. The problem is that our eye interprets that leftmost node as the root, when in fact there is no root. If we really want to display the tree in an unbiased fashion, we need to draw it in the unrooted or **Radiation** format (**Figure 6.7**).

An unrooted tree has no "direction" associated with it; we cannot say that one node is descended from another because we know nothing about the order of descent and we can trace a path from any interior node to any tip sequence. In contrast, in a rooted tree, from any particular interior node we can only trace paths to a *subset* of the tip sequences (because we cannot move backward toward the root). That subset of a rooted tree is called a clade, and the members of a clade are descended from the ancestor represented by that particular node. Chapter 7 discusses how to go about determining the root of an unrooted tree.

If we remove any branch from an unrooted tree we "split" the taxa into two mutually exclu-

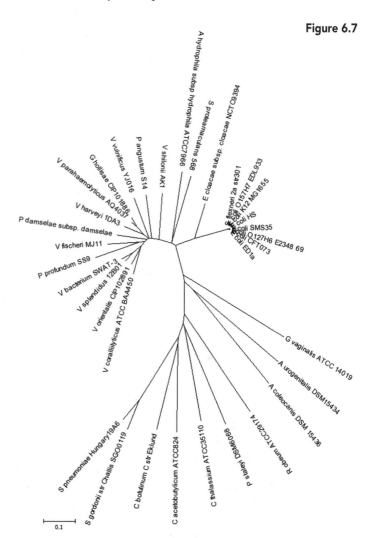

Figure 6.7

sive groups (or splits). One split includes the taxa at one end of the removed branch, while the second split includes the taxa at the other end of the branch. We will consider splits in detail in Chapter 15. For the moment it is sufficient to keep in mind that *clades* refer to rooted trees and *splits* refer to unrooted trees—and also to keep in mind that all of the methods in this book estimate unrooted trees.

Rooted trees are so integral to our concept of evolution that the terminology of phylogenetics often unconsciously assumes that all trees are rooted. For instance, a *strictly bifurcating tree* is defined as one in which each internal node has *exactly* two branches descending from it—even if the tree is unrooted. A *polytomy* is a node with *more than* two branches descending from it—even if the tree is unrooted and there is no descent involved. I use that slightly sloppy terminology because the alternatives are really awkward.

Estimating the Reliability of a Tree

One of the most important things to understand about the phylogenetic trees that we estimate is that they are almost certainly wrong. Even if we ignore branch lengths and consider only topology, there are about 2×10^{182} possible topologies for a tree of 100 sequences (see *Learn More about Phylogenetic Trees*, p. 72). Tree reconstruction methods such as Neighbor Joining attempt to select the one tree that most accurately represents the historical branching order of the sequences. The estimated tree is highly unlikely to be the true tree. However, that tree is *the best estimate we can make* given the assumptions of the method, the model that has been chosen, and the implementation of that method. We cannot know the true tree, but we can do our best to estimate a tree that is reasonably close to the true tree. Given that the trees we obtain are estimates, we need some way to assess the *reliability* of those estimates.

The most widely used method for estimating reliability of phylogenetic trees is the **bootstrap method** (see *Learn More about Estimating the Reliability of Phylogenetic Trees*, p. 83), although other methods, such as Bayesian posterior probabilities, are becoming more widely used. It is important to understand that these methods estimate *reproducibility*, not accuracy. In the case of bootstrapping, the reproducibility with which splits exist in trees is based upon subsamples of the data. Fortunately, both bootstrap and posterior probabilities are known to be conservative estimates. Simulation studies, in which the true tree is known and can be compared with the estimated tree, have shown that both bootstrap and posterior probabilities underestimate the probability that a split is correct.

To perform a bootstrap test, change **Phylogeny Test** in the **Analysis Preferences Window** from **None** to **Bootstrap Method** (red arrow in **Figure 6.8**). A new option, **No. of Bootstrap Replications** (black arrow in Figure 6.8), will appear with a default setting of 500. You should set the number of replications to at least 100, and 2000 is preferable.

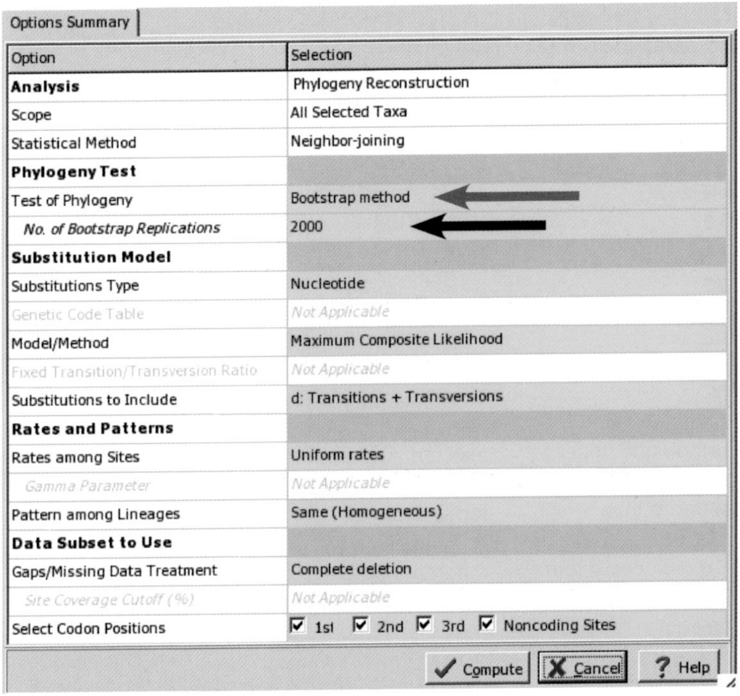

Figure 6.8

Options Summary	
Option	**Selection**
Analysis	Phylogeny Reconstruction
Scope	All Selected Taxa
Statistical Method	Neighbor-joining
Phylogeny Test	
Test of Phylogeny	Bootstrap method ←
No. of Bootstrap Replications	2000 ←
Substitution Model	
Substitutions Type	Nucleotide
Genetic Code Table	Not Applicable
Model/Method	Maximum Composite Likelihood
Fixed Transition/Transversion Ratio	Not Applicable
Substitutions to Include	d: Transitions + Transversions
Rates and Patterns	
Rates among Sites	Uniform rates
Gamma Parameter	Not Applicable
Pattern among Lineages	Same (Homogeneous)
Data Subset to Use	
Gaps/Missing Data Treatment	Complete deletion
Site Coverage Cutoff (%)	Not Applicable
Select Codon Positions	☑ 1st ☑ 2nd ☑ 3rd ☑ Noncoding Sites

✓ Compute   ✗ Cancel   ? Help

LEARN MORE ABOUT

Estimating the Reliability of Phylogenetic Trees

How well can you trust the tree you have just estimated? It depends on what features of the tree are important to you. Most of the time, "reliability" refers to the *topology*, or branching order, of a tree, not to the lengths of the branches. In essence, reliability is measured as *the probability that the members of a given clade are always members of that clade*.

An experimental scientist who wants to test the reliability of a conclusion repeats the experiment with independent data. Since the data in this case are the sequences themselves, and sequences are what they are, there seems to be little point to repeating the data unless we just want to test the reliability of the sequencing. We might repeat the alignment, but unless we change the gap parameters we will simply regenerate the same alignment.

Phylogeneticists use a sampling method called **bootstrapping** that pseudorepeats the collection of data as a method to estimate the reliability of the tree.

Consider the following alignment:

```
1234567890
AATGGGATTT
CCCGGGGCCG
AATGGAGATT
TATGGCGGTT
TGTGGGCATT
```

A random site (i.e., a column) is taken from the alignment and used as the first site in a pseudoalignment. Another random site is taken and used as the second site in the pseudoalignment, and the process is continued until the pseudoalignment contains the same number of sites as the original alignment. Sampling of the original alignment is with replacement, which means that the same site may be placed in the pseudoalignment more than once, or may not appear in the pseudoalignment at all. The pseudoalignment might look like this:

```
4836024951
GTTGTAGTGA
GCCGGCGCGC
GATATAGTGA
GGTCTAGTGT
GATGTGGTGT
```

A tree is then constructed from the pseudo-alignment using the same method and under the same parameter settings used to estimate the original tree. The original tree is then compared with the new tree. For every clade in the original tree, a score of 1 is assigned if that clade is present in the new tree; a score of 0 is assigned if the clade is not present in the new tree. That process constitutes *one bootstrap replicate*. The score for each clade is recorded and the next bootstrap cycle is initiated.

Obviously, a bootstrap replicate requires slightly longer than the time required to estimate the original tree. Typically 100–2000 bootstrap replicates are used to estimate tree reliability, but that depends somewhat upon the time required for each replicate. In the end, a tree is printed showing, for each clade, the number of times (or fraction of times, depend-

ing on the program) that the clade occurred in the bootstrap replicates. We can be pretty confident in a clade with a 90% bootstrap value, and have very little confidence in a clade with a 25% bootstrap value. Interpretation of the reliability scores depends entirely on how important the details of a particular clade are to the point you are trying to make with the tree.

Because both NJ and Parsimony are relatively fast methods, you can almost always run a bootstrap analysis on trees constructed by those methods, and it is reasonable to do 2000 bootstrap replicates. Maximum Likelihood, on the other hand, can take a very long time to execute 2000 bootstrap replicates. You are more likely to find yourself using the lower limit of 100 replicates.

Bayesian Inference (BI) takes an entirely different approach to assessing tree reliability. Instead of bootstrapping with pseudodata, Bayesian Inference permits directly counting the fraction of times a clade occurs among the trees sampled (see *Learn More about Bayesian Inference*, p. 141). Bayesian Inference thus requires no additional time to estimate tree reliability. MrBayes is also unusual in that it allows the user to present both the reliability of clades and branch lengths on a consensus tree—which amounts to having your cake and eating it too.

For a good and very readable discussion of bootstrapping, see Li 1997. ◄

The higher the number of replications the longer the test will take, but NJ is so fast that it is reasonable to set the number to 2000 (using numbers larger than 2000 offers little gain). Once the number of replications and the other parameters are set, click **Compute** to start the bootstrap analysis of tree reliability. A window with a progress bar shows how the analysis is proceeding. It typically takes 1–2 minutes for 2000 replications.

When the tree appears, it displays numbers next to each node (**Figure 6.9**). The numbers are the *bootstrap percentages*—that is, the percent of bootstrap replicates in which both splits generated by removing a branch in the real tree were also generated by removing that branch in the replicates. Although the number is typically placed adjacent to a node, it is the **branch support** for the *branch* connecting to that node and is often referred to as the **bootstrap value**.

Figure 6.9

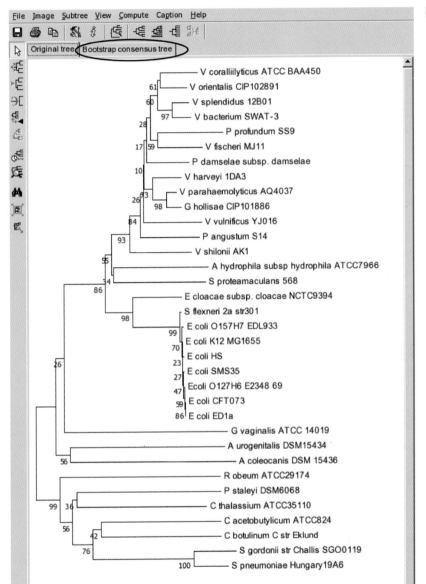

Click the **Bootstrap Consensus** tab (circled in Figure 6.9) to display the bootstrap consensus tree. The bootstrap consensus tree may or may not be identical to the original tree. In this format, known as the **phylogram** format, branches are drawn so that the line lengths are proportional to the branch lengths. That format can make it difficult to see which bootstrap label applies to which node.

The **cladogram** format (**Figure 6.10**) allows you to see the bootstrap support for each branch more clearly because the lengths of the branch lines in the drawing are not proportional to the branch lengths. To display the tree in the cladogram format, choose **Topology Only** from the **View** menu (**Figure 6.11**) or click the **Display Only Topology** button (circled in Figure 6.10).

Figure 6.10

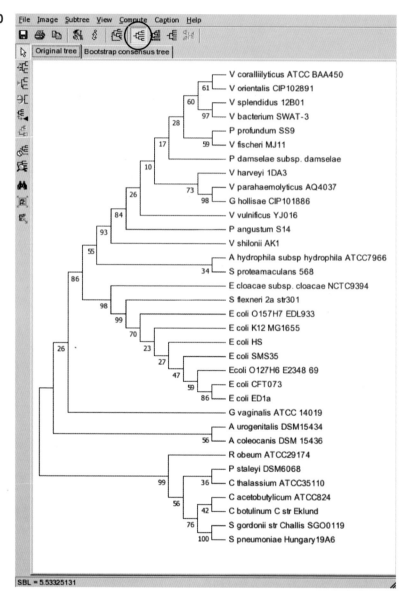

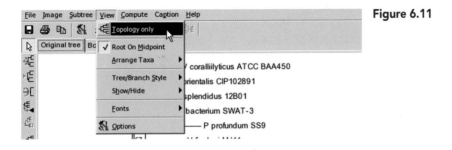

Figure 6.11

The bootstrap values reflect the uncertainty in the branching order at each point. We don't take branches with less than 70% support very seriously. Figure 6.10 shows the **strict consensus** tree—the most frequent branching order at each node. Notice that several branches have less than 50% support, meaning that we *really have no decent idea* what the branching order is. A common way to make that uncertainty clear is to collapse branches with <50% support into **polytomies**—nodes from which more than two branches descend. To show the resulting **majority rule tree**, choose **Condensed Tree** from the **Compute** menu and in the resulting **Options** window (**Figure 6.12**) set the **Cut-off Value for Consensus Tree** to 50% and click **OK**.

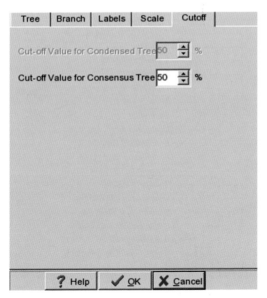

Figure 6.12

Figure 6.13 shows the majority rule tree. Notice that all of the bootstrap values are now ≥50%, but that many nodes have multiple branches descending from them.

Figure 6.13

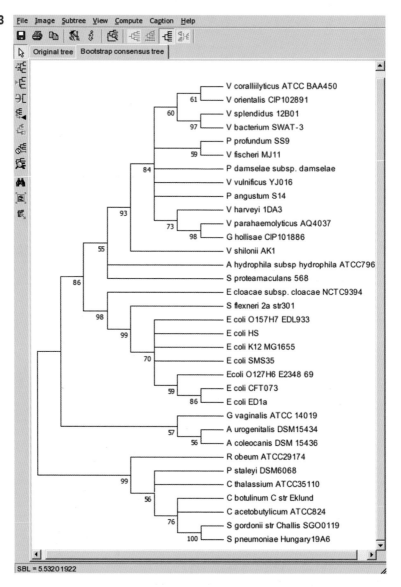

The final order of business is to choose **Save** from the **File** menu to save the trees in MEGA's .mts format. The tree drawing can be printed from the same **File** menu.

Download Chapter 6: LargeData.meg

You can reproduce this exercise with the LargeData.meg file if you wish. **Figure 6.14** shows the majority rule bootstrap consensus tree of the LargeData set (see Chapter 5).

Figure 6.14

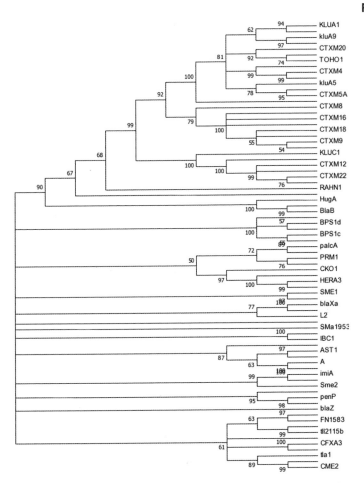

What about Protein Sequences?

We are so accustomed to inferring protein sequences from DNA sequences that we may forget that there was a time when people actually sequenced proteins directly. If no nucleotide coding sequences exist for some of your protein sequences, you will need to base your NJ tree on the protein sequences themselves. The only difference is that when choosing the **Substitution Type** (see Figure 6.4), you must change **Nucleotide** to **Amino Acid** and then choose an amino acid model instead of a nucleotide model. Of the amino acid models you will choose from, the Poisson correction model is roughly equivalent to Jukes-Cantor for nucleotides: it corrects for multiple hits. The Dayhoff and JTT models also correct for multiple hits, but include substitution rate matrices based on observed proportions of amino acid substitutions in large sets of proteins. The Dayhoff matrix dates from 1979. The JTT matrix is a 1992 update of the Dayhoff approach with a much larger set of proteins. I suggest using the JTT matrix model for protein-based NJ trees.

CHAPTER 7

Drawing Phylogenetic Trees

Figures 6.6 and 6.7 do not show phylogenetic trees, they show *drawings* of phylogenetic trees. A tree is an abstract structure consisting of branches and nodes; it is analogous to the X–Y coordinates of a set of points. We could describe that set by listing the coordinates, but most often we choose to draw a graph that makes it easier for the audience to understand the relationships among the coordinate points. For the same reason, we draw a tree to make it easier to visualize the historical relationships among the sequences of interest. What we saved in Chapter 6 was not a drawing; it was a *description* of the sequence relationships saved in a particular file format (.mts) from which MEGA can quickly redraw the tree.

After carefully selecting the sequences that will appear on the tree, downloading them, aligning them, choosing a phylogenetic method and evolutionary models, etc., the act of drawing the tree is virtually instantaneous. The temptation is to simply accept the drawing. However, that drawing is the means by which you will interpret your tree in order to understand some biological problem. Of equal importance, it is the means by which you will convey that understanding to your audience. The appearance of the tree is therefore as critical, and as deserving of your thoughtful attention, as any other step in the estimation of a phylogenetic tree.

Just as we can draw a graph in a variety of ways to represent the same information, we can draw phylogenetic trees in a variety of ways. This chapter focuses on the different ways to draw trees that will allow you to make the information as clear as possible and focus attention on the point you are trying to make with the tree.

Changing the Appearance of a Tree

Figure 7.1 shows the LargeData tree in the familiar **rectangular phylogram** format in MEGA's **Tree Explorer** window. In a *rectangular* format, interior nodes are represented by vertical lines and branches are represented by horizontal lines. The *phylogram* format means that the horizontal branch lines are drawn so that their lengths are proportional to the number of substitutions per site; a scale at the bottom of the drawing shows the number of substitutions per site that are represented by a branch line of a particular length.

Figure 7.1

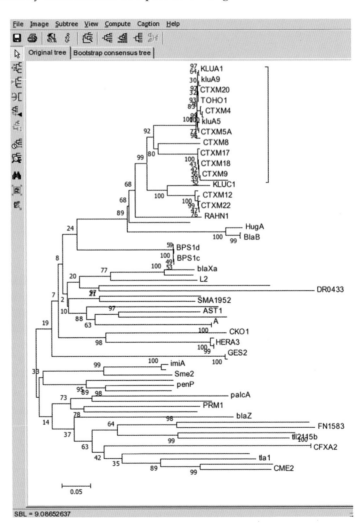

One advantage of the rectangular phylogram format is that it makes branch lengths dramatically obvious. On the other hand, when there are some long branches, it can be hard to see the branching order of sequences that are related by short branches (e.g., within the bracketed clade). It is also impossible to tell which bootstrap percentage belongs to which node. In such a case it can be helpful to use the alternative **rectangular cladogram** format in which branch lines are not drawn proportionally to branch lengths. To display the tree in the cladogram format, click the **Display Topology Only** button in the **Tree Explorer** window (red arrow in **Figure 7.2**).

Figure 7.2

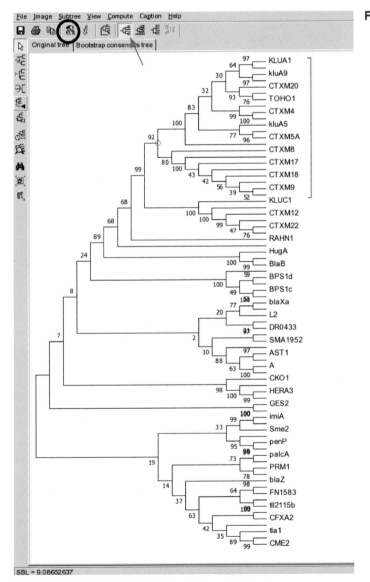

The Options dialog

The problem with Figure 7.2 is that we can no longer see how long the branches are. To remedy that problem, click the **Options** icon (circled in Figure 7.2) to display the **Tree Options** dialog (**Figure 7.3**). This innocuous little window gives you enormous detailed control over the appearance of the tree. When you are unsure about how to modify the appearance of a tree, look here first. Click the **Branch** tab, then tick the box that says **Display Branch Length** (**Figure 7.4**).

Figure 7.3

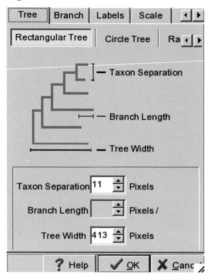

Figure 7.4

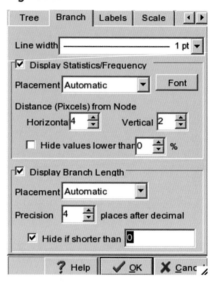

Notice that the **Branch** tab also allows you to control whether the bootstrap values are displayed (**Display Statistics/Frequency**), and for both **Statistics/Frequency** and **Branch Lengths** it allows you to choose the placement of the label (**Automatic, Above Branch,** or **Below Branch**). For **Display Statistics/Frequency** you can set the offset from the node both horizontally and vertically, and for **Display Branch Lengths** you can set the number of decimal places. You can change the **Font** for both the branch and node labels, although you cannot set them independently. Finally, you can set the **Line Width** (line thickness) for the branches and nodes.

Now we see branch lengths printed below each branch, with bootstrap percents printed to the left of each node (**Figure 7.5**). While that tree does indeed display all the information of interest, it is so crowded that it is difficult to look at, much less interpret. We can improve things somewhat by re-sizing the window to make it wider, then clicking the **Fit Tree to Screen** button (see Figure 7.10). We can improve things further by using the **Labels** function in the **Options** dialog to set the font size down to 6 points, but the appearance is still not wonderful. Sometimes there is no perfect solution, so you must decide exactly what information is important to display. Given your purpose(s) for presenting this

Figure 7.5

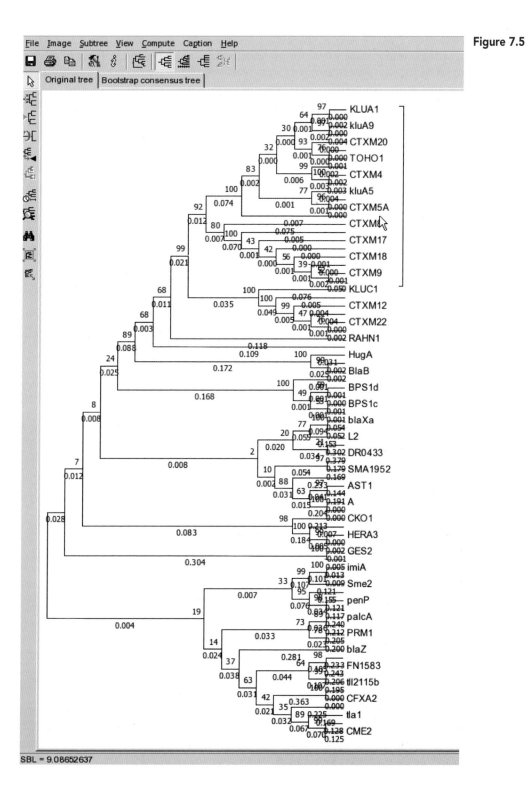

SBL = 9.08652637

tree, does it really matter what the branch lengths are? If not, don't bother to display branch lengths, but simply use the tree as it appears in Figure 7.2. Are all of the bootstrap values so high that you can simply state "Bootstrap support for all clades was >95%"? If so, the phylogram format in Figure 7.1 would do very nicely. The point is that there is no "right" solution to the problem; the appropriate solution is situational and requires some thought.

Branch styles

Figure 7.6 shows the majority-rule bootstrap consensus tree for the SmallData in the **Traditional Rectangular** format. The **Tree Branch Style** button at the top of the screen permits choosing other styles as well.

Figure 7.6

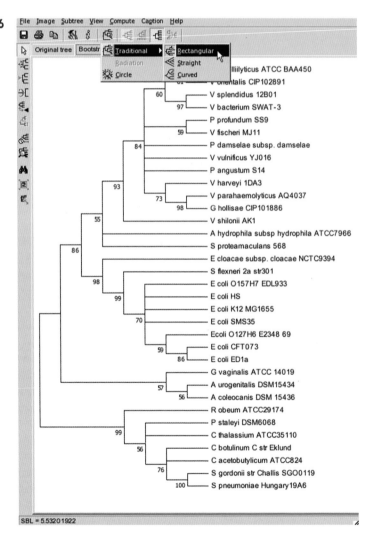

Choosing the **Straight** (sometimes called slanted) style displays nodes as the intersection of branches (**Figure 7.7**). This straight style makes it clear that branches lead from a common ancestor. It is especially helpful when there is a polytomy (more than two branches descending from a single node).

Figure 7.7

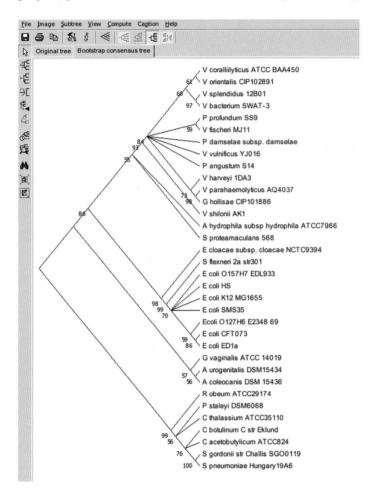

Some people find the **Circle** style (**Figure 7.8**) more appealing than a traditional branching tree format.

Figure 7.8

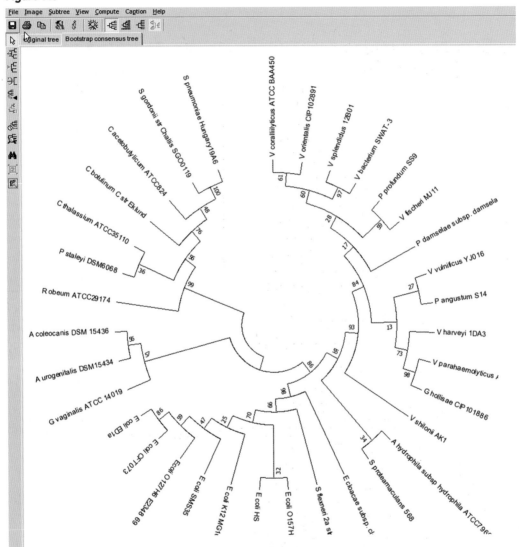

The **Radiation** format displays a tree in its unrooted condition (see Figure 6.7 and pp. 80–82). The point is that, as dissimilar as the different styles look, they show exactly the same information. It is up to you to choose the style that you think will convey the important aspects of that information most effectively.

Fine-tuning the appearance of a tree

When the LargeData tree is first displayed it is too large to see the entire tree (**Figure 7.9**).

Figure 7.9

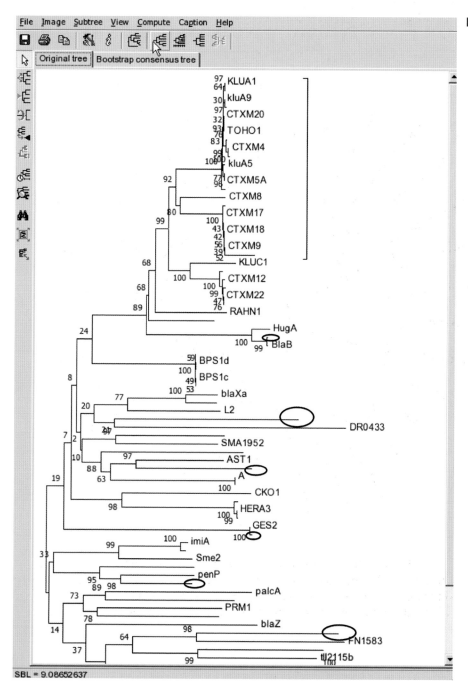

You already know how to scale the tree to fill the available window space by clicking the **Fit Tree to Screen** button, but the **Tree** tab of the **Options** dialog offers finer control over tree size by allowing you to individually adjust **Taxon Separation**, **Branch Length**, and overall **Tree Width** (see Figure 7.3).

In addition, the icons displayed along the left edge of the **Tree Explorer** window control a host of tools for modifying the appearance of the tree. **Figure 7.10** labels these icons for discussion purposes.

Figure 7.10

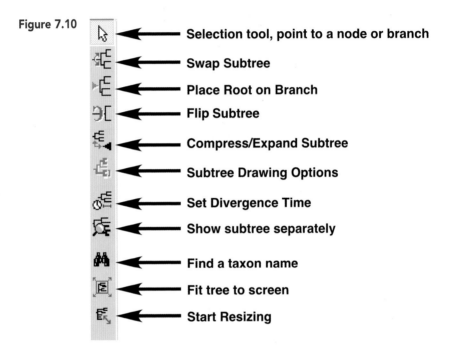

Selection tool, point to a node or branch

Swap Subtree

Place Root on Branch

Flip Subtree

Compress/Expand Subtree

Subtree Drawing Options

Set Divergence Time

Show subtree separately

Find a taxon name

Fit tree to screen

Start Resizing

Start Resizing lets you drag with the mouse to change the vertical or horizontal size as you watch. In Figure 7.9, there are several branches (circled) that have no taxon label. Why? Because when taxon labels overlap and become unreadable, MEGA simply doesn't display one of them. Using the **Start Resizing** tool we can expand the drawing vertically so that all taxon names are displayed (**Figure 7.11**). Of course, this expansion comes at the cost of seeing even less of the tree.

Figure 7.11

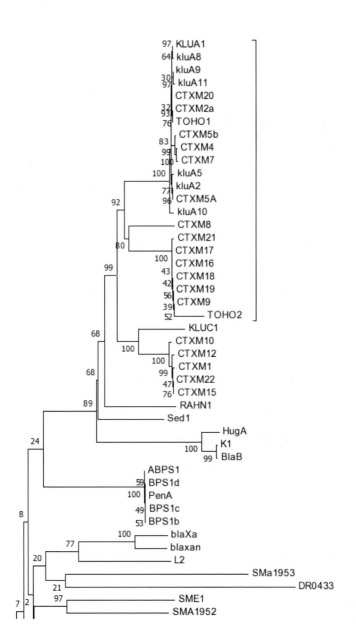

The **Selection tool** lets you click on a node or branch to select it for a further action. When you click on a node it is identified by a red diamond. When you click on a branch it is identified by a green rectangle. The **Swap Subtree** and **Flip Subtree** tools allow you to rotate a clade around the selected node. **Figure 7.12A** shows the original tree with the node selected (red diamond, cursor arrow). **Figures 7.12B** and **7.12C** show the results of **swapping** and **flipping** the clade respectively. Keep in mind that all three views are exactly the same tree; all that has changed is the order in which the branches are written within a clade.

Figure 7.12

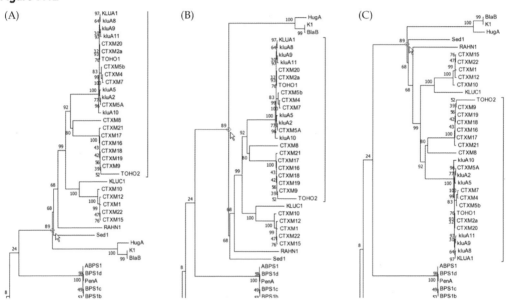

Subtrees

In **Figure 7.13**, the set of sequences labeled CTXM constitutes a clade that is descended from the node indicated by the black arrow. This clade breaks down further into three distinct clades that are separated by moderately long branches, but within each of those clades the sequences have diverged very little. Indeed, it is almost impossible to see the branching orders within those clades.

Figure 7.13

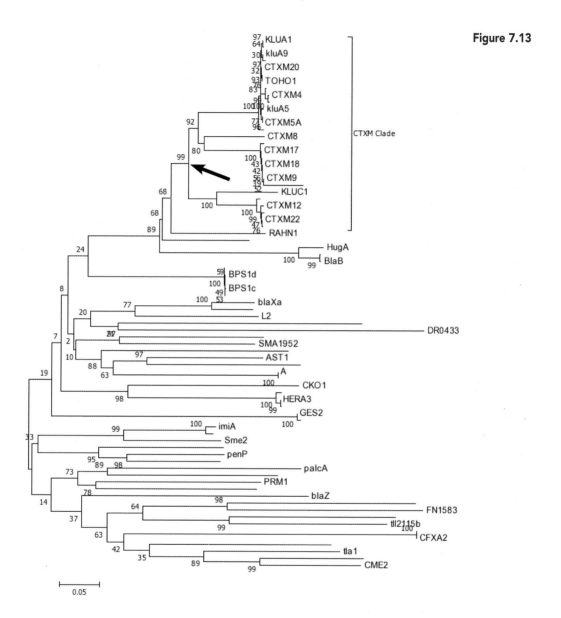

Figure 7.14

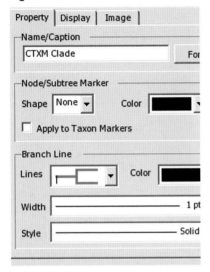

The **Compress/Expand Subtree** tool (see Figure 7.10) allows you to display the CTXM clade as a single subgroup. Click that tool to activate it, then click on the branch leading to the node that is common to the CTXM clade. A dialog box (**Figure 7.14**) requires you to name the subgroup.

The **Subtree Drawing Options** window then allows you to set a variety of properties, including the color of the subtree triangle. In this case, I named the subgroup CTXM clade. The result is the tree shown in **Figure 7.15**.

Clicking on the CTXM clade branch again with the **Compress/Expand Subtree** tool reverses the process and restores the original appearance of the tree—except that the clade is enclosed in a bracket and the bracket is labeled with the name you chose. Indeed, that is how I created the bracket in Figure

Figure 7.15

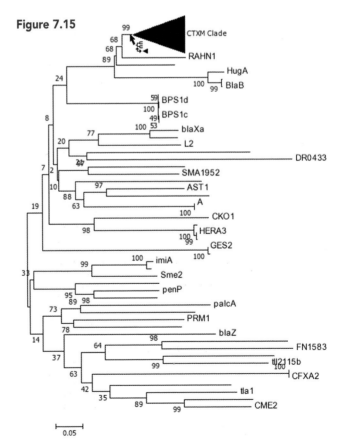

7.13. To remove that label and bracket—that is, to completely eliminate the subtree—choose **Draw Options** from the **Subtree** menu, then choose **Clear All Subtree Drawing Options** (**Figure 7.16**).

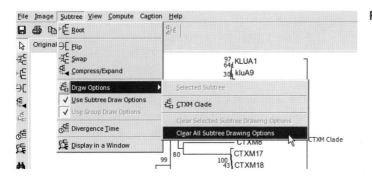

Figure 7.16

You can also invoke that dialog by selecting the branch leading to the subtree and clicking the **Subtree Drawing Options** button (see Figure 7.10).

The **Show Subtree Separately** tool allows you to display the subgroup that appears as a black triangle in Figure 7.15 as a separate tree. Click the tool on the branch that leads to the subgroup to display the subtree in a separate window (**Figure 7.17**).

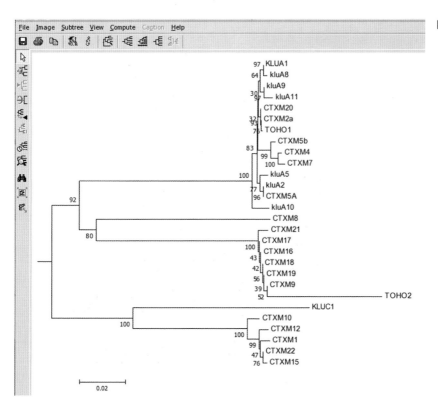

Figure 7.17

The subtree now allows you to see the branching order and the branch lengths in much more detail than was possible in Figure 7.13. Notice that in the drawing of the subtree in Figure 7.17 all of the branches are labeled with taxon names, whereas in Figure 7.13 (the complete tree), some taxon names are absent.

The subtree feature is not only useful for seeing details that are otherwise obscured, it can also help you deal with very large trees. For instance, a tree with 100 or more sequences may break down into three or four distinct clades. It may be better to show one tree in which each clade is displayed as a subgroup triangle, then show the details of each subtree in separate figures. Then the audience can see all of the necessary details while still being able to visualize the relationships among the clades.

Rooting a Tree

As pointed out in Chapter 6, the SmallData tree in Figure 6.6 only *appears* to be rooted because of the way rectangular trees are necessarily drawn. It is in fact an unrooted tree. Usually, however, we really are interested in having a rooted tree—one that shows the direction of evolution—so that we can make inferences about the order of descent.

The rectangular format is obtained by selecting a branch of an unrooted tree and placing an interior node—the root—within that branch. Some tree-drawing programs simply place the apparent root in the branch that leads to the first sequence listed in the alignment. MEGA improves upon that by placing the apparent root at the midpoint of the NJ tree, in the branch located midway between the two most distant sequences. If the rate of evolution is roughly constant along all branches, then this midpoint rooting will have placed the root correctly. More often, however, midpoint rooting places the root incorrectly, so it should not be trusted. Indeed, the apparent root in Figure 6.6 is placed incorrectly. So how can we find the root of the tree?

There is not sufficient information in the sequences themselves to accurately place the root, so we need additional outside information. That outside information is in the form of an outgroup. An **outgroup** is defined as *one or more sequences that are more distantly related to the ingroup sequences than the ingroup sequences are to each other*. For the tree in Figure 6.6, I know that all the species from *G. vaginalis* down through *S. pneumoniae* are Gram positive, while the species appearing above *G. vaginalis* are all Gram negative. I also know that the Gram positive bacteria diverged from their common ancestor with the Gram negative species about 2.2 billion years ago. Thus I can be confident that the Gram positive species are a legitimate outgroup to the Gram negative species.

The **Place Root on Branch** tool (see Figure 7.10) is used to place the root within a desired branch (the branch shown in red in Figure 6.6). After clicking on that branch with the tool, the tree looks like **Figure 7.18**. The leftmost node, now the root, leads to two groups: the Gram negative ingroup at the top and the Gram positive outgroup at the bottom. Compare the rooted tree in Figure 7.18 with the unrooted trees in Figure 6.6 and 6.7.

Figure 7.18

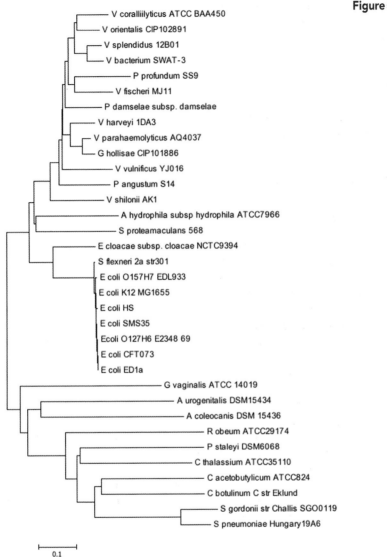

It is very important not to confuse the *appearance* of a rooted tree with an actual rooted tree. When you see a tree in a paper or in a seminar, unless the author states that the tree is rooted, and how it was rooted, assume that it is actually an unrooted tree. This means that any conclusions based upon the order of descent of the taxa or sequences in that tree should be dismissed as unreliable.

Finding an outgroup

What if there is no obvious outgroup among the sequences on the tree? Well, then you can *add* some outgroup sequences to the alignment. Suppose you are dealing with a set of sequences, all of which come from mammals. You could look for one or more homologous sequences from birds. You know that birds diverged from mammals before mammals began to diverge from each other, so those sequences will certainly be outgroup sequences. The tricky part is that you must also be sure that any bird sequences used are homologs of the mammalian sequences. In other words, *outgroup sequences must have diverged before the ingroup sequences diverged from each other, but they must not have diverged so much that homology is not detectable.* (See Chapter 3 for how to use the **BLAST2seq** application to detect homology, and Chapter 12 for how to determine whether outgroup sequences are sufficiently homologous to permit good alignment.)

A key aspect of defining one or more sequences as an outgroup is that we must use information that is external to the sequences themselves. You cannot, for instance, decide that a sequence constitutes an outgroup just because it is the most genetically distant of the sequences. Genetic distance is information that is *internal* to the sequences; i.e., it is based on properties of the sequences themselves.

Saving Trees

Having gone to the trouble of editing the tree drawing so that it is now looks the way you want it to, it is important to save that tree. There are two distinct ways to save a tree: (1) as a tree description, and (2) as a tree image.

Saving a tree description

The tree description is saved in MEGA's .mts format by clicking the **Disk** icon at the top left of the window, or by choosing **Save** from the **File** menu. The .mts format is used only by MEGA, so if you want the tree description to be used by another program you will need to choose **Export Current Tree** from the **File** menu and save the description in the Newick format.

Newick (named for the restaurant in which a group of systematists devised the format) is used by most tree-drawing programs. Indeed, MEGA can even draw trees from Newick description files that have been written by other phylogenetics programs; simply choose **Display Newick Trees** from the **User Trees** menu of the main MEGA window. The resulting file will have the extension .nwk. The documentation for the Phylip programs Drawtree and Drawgram (http://evolution.genetics.washington.edu/phylip/doc/draw.html) includes a nice description of the Newick format.

Saving a tree image

The tree drawing itself can be saved by choosing **Save As** from the Image menu of the **Tree Explorer** window. MEGA offers three graphics formats for saving tree images: Enhanced Metafile (EMF), PDF, and TIFF.

Files saved in the EMF format can be opened by most Windows graphics programs (including PowerPoint) in order to place additional elements—such as arrows, labels, and so forth—on the tree. Because EMF is a vector graphics format, the drawing can be manipulated without loss of image quality.

PDF (Portable Document Format) files are widely accepted and can be opened and printed by Mac's Preview program. Although not editable in itself, PDF is a vector format that can be visualized and modified by Adobe Illustrator. PDF is the preferred format for saving trees in MEGA for Mac.

The TIFF format is accepted by most journals for publishing figure images, so if no manipulation of the drawing is required, and if a journal requires TIFF images, you can save the image in that format. Because TIFF is a bitmap format and thus difficult to manipulate, it is a good idea to first use the various tree-drawing options in MEGA to make the tree exactly the way you want it to appear in print.

Captions

MEGA includes a **Caption** command that creates an automatic caption for every analysis that MEGA performs. To generate the caption for a phylogenetic tree, choose the **Caption** menu in the **Tree Explorer** window (see Figure 7.1). The resulting Caption window describes the exact conditions under which the tree was estimated, including bootstrap setting, and provides relevant citations. This information can be saved, printed, or copied to the clipboard.

MEGA's caption feature is enormously helpful when it comes time to write a paper. Users are advised to save a caption each time they create a tree. A few weeks or months later (or whenever you get around to writing the paper), it can be surprisingly difficult to recall the details of the analysis. At the same time, users are advised to modify the MEGA caption for use in a publication. Think of the Caption feature as a good starting point for a figure legend or a text description of a tree.

CHAPTER 8

Parsimony

Parsimony, or minimum change, is based on the assumption that the most likely tree is the one that requires the fewest number of changes to explain the data—protein or nucleotide sequences—in the alignment (see *Learn More about Parsimony,* below*).* Instead of the **Models** option used in Neighbor Joining, MEGA uses an **MP Search Method** option to implement Parsimony. That is because Parsimony does not use specific evolutionary models of the nucleotide substitution process to estimate trees (and thus in this method there is no difference in the way MEGA handles protein versus nucleotide sequence alignments). Instead of choosing a model, we must choose the method by which the program searches for the minimum number of steps (i.e., the most parsimonious tree).

LEARN MORE ABOUT

Parsimony

The basic premise of Parsimony is that taxa sharing a common characteristic do so because they inherited that characteristic from a common ancestor. When conflicts with that assumption occur (and they often do), they are explained by reversal (a characteristic changed but then reverted back to its original state), convergence (unrelated taxa evolved the same character independently), or parallelism (different taxa may have similar properties that predispose a characteristic to develop in a certain way). These explanations are gathered together under the term **homoplasy**. Homoplasies are regarded as "extra" steps or hypotheses that are required to explain the data. More formally, Parsimony assumes that a character is more likely to be common to two taxa because it was inherited from a common ancestor than it is to be common because of homoplasy.

Parsimony operates by selecting the tree or trees that minimize the number of evolutionary steps, including homoplasies, required to explain the data. Parsimony, or minimum change, is the criterion for choosing the best tree.

For protein or nucleotide sequences alike, the data are the aligned sequences. Each site in the alignment is a character, and each character can have different states in different taxa. Not all characters are useful in constructing a Parsimony tree. Invariant characters—those that have the same state in all taxa—are obviously useless and are ignored by the method. Also ignored are characters in which a state occurs in only one taxon.

An algorithm is used to determine the minimum number of steps necessary for any given tree (i.e., any given branching order) to be consistent with the data. That number is the score for the tree, and the tree or trees with the lowest scores are the most parsimonious trees.

The algorithm is used to evaluate a possible tree at each informative site. Consider a set of six taxa, conveniently named 1–6. At some site (character) in the alignment, the states of that character are:

$$1=A$$
$$2=C$$
$$3=A$$
$$4=G$$
$$5=G$$
$$6=C$$

There are 105 possible unrooted trees of six taxa. We will pick the unrooted tree in Figure 1 as our example, but all 105 will be evaluated by the computer.

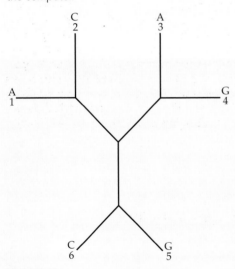

FIGURE 1

If we root that tree at taxon 1, we get the tree in Figure 2.

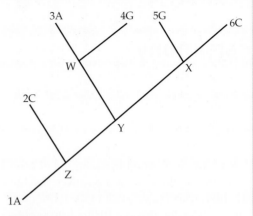

FIGURE 2

The algorithm starts at a tip and moves to the interior node that connects to another tip. If the two tips have the same state, the algorithm assigns that state to the node; if they do not, it assigns an "or" state. Thus node W is assigned the state A or G, and node X the state G or C. Node Y connects nodes W and X. Because the states at nodes W and X both include G, node Y is assigned the state G. Node Z is assigned the state C or G (Figure 3).

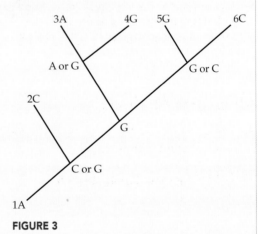

FIGURE 3

Once the root has been reached, the algorithm proceeds back up from the root toward the tips. Because node Z does not include the state at the node that is ancestral to it (taxon 1), its assignment is arbitrary. Assume that it is assigned state G. Node Y is already assigned, so the algorithm moves to node W. Node W is assigned G because that assignment does not require a change from the node that is ancestral to it. Similarly, node X is assigned state G (Figure 4).

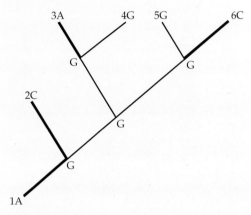

FIGURE 4

Each branch along which the state changed, indicated by thick branches, is counted. This tree has four changes. If node Z had been assigned state C instead of state G, the resulting

tree—which also has four changes—would be as shown in Figure 5.

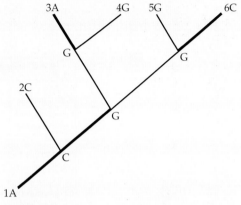

FIGURE 5

The other possible rootings of the tree are considered in the same way, and if a different rooting of the tree produces fewer changes, that lower number is the score for that site. The Parsimony program evaluates the tree for each informative site, then adds up the changes to calculate the minimum number of changes for that particular tree. As it works its way through the various possible trees, the program keeps track of the tree (or trees) with the lowest scores.

There is a nice discussion of Parsimony in Li 1997, pp. 112–115, and a more detailed discussion in Swofford et al. 1996, pp. 415–425.

MP Search Methods

To estimate a tree by Maximum Parsimony (MP), open the SmallData.meg file in MEGA's main window.

 Chapter 6: SmallData.meg

Choose **Construct/Test Maximum Parsimony Tree(s)** from the **Phylogeny** menu. The resulting **Analysis Preferences** dialog, seen in **Figure 8.1**, looks familiar but it is somewhat different from the NJ dialog. Set the **Gaps/Missing Data** option exactly as for Neighbor Joining: if there are many gaps choose **Partial Deletion**, otherwise choose **Complete Deletion**. Then Click the yellow **MP Search Method** field to display the tree search options.

Figure 8.1

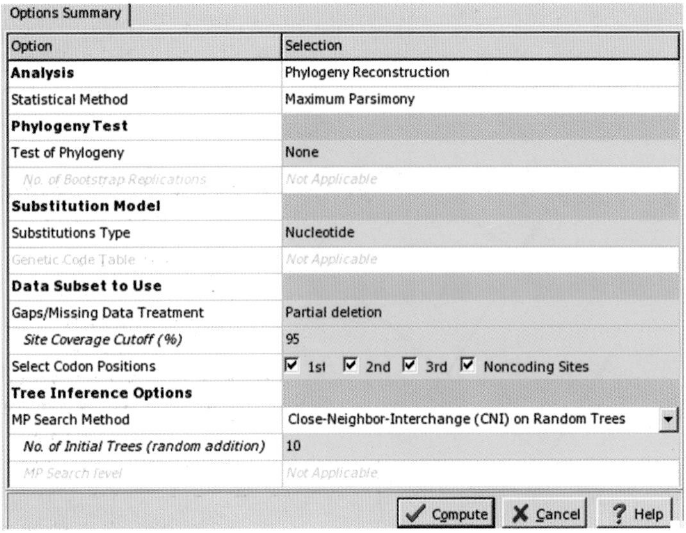

Three basic search methods are available: **Close-neighbor-interchange** (**CNI**), **Min-Mini Heuristic**, and **Max-mini Branch-&-bound** (**Figure 8.2**). The latter two methods are very slow and I do not recommend them.

Figure 8.2

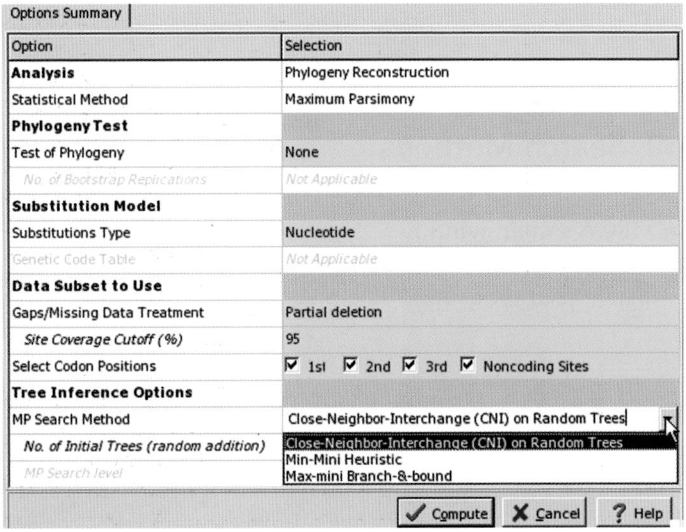

Within Close-Neighbor-Interchange (CNI) you can choose the number of initial trees for random addition. I recommend selecting CNI with the default setting of 10 trees (see Figure 8.1) because it is considerably faster than the Min-Mini method. The Max-Mini Branch-&-Bound method is guaranteed to find the most parsimonious tree(s), but it is *very* slow and should not be used on data sets with more than 15 sequences (taxa).

Figure 8.3 shows the Parsimony tree that results from using the SmallData. meg file with Partial Deletion of gaps and CNI with 10 initial trees. You will notice that the tree is in the cladogram format and that the **Show Topology Only** button is highlighted. *Do not touch that button* (at least, not yet)! The reason only the topology is shown is because branch lengths have not been calculat-

Figure 8.3

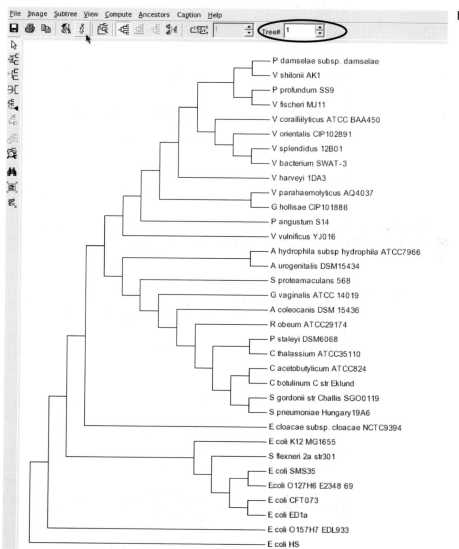

ed. Calculating branch lengths can take a *very* long time, and that calculation will begin if you click the button (see pp. 117–118).

Multiple Equally Parsimonious Trees

Figure 8.3 shows one parsimony tree. Clicking the i (for information) button (indicated by the cursor arrow) shows the information dialog (**Figure 8.4**), which tells us we are in fact looking at tree #1 of 6 different trees.

Figure 8.4

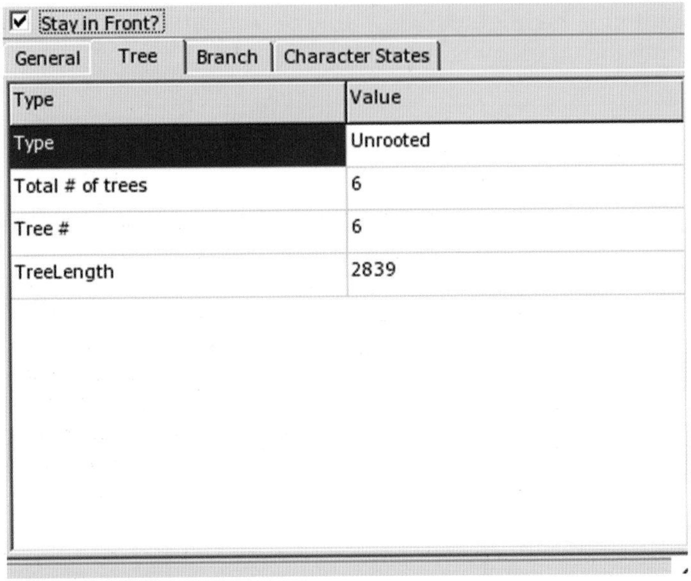

☑ Stay in Front?	
General **Tree** | Branch | Character States |	
Type	**Value**
Type	Unrooted
Total # of trees	6
Tree #	6
TreeLength	2839

In the **Tree#** box at the top of the **Tree Explorer** window (circled in Figure 8.3), we can click the "up" arrow to see additional trees. **Figure 8.5** shows tree #4, which is different from tree #1. This difference is not an error or a failure in the program; it is an inherent property of the Maximum Parsimony method. As an analogy, if you were driving from New York to Los Angeles and seeking the shortest route, you might discover several different routes that are the same distance. In a similar sense, there can be many trees that are different but require the same number of steps to explain them. This is one such case.

You will also notice that not only is the tree in Figure 8.5 in the cladogram (topology only) format, it is not even midpoint-rooted as the NJ tree in Figure 6.6 was. Both differences stem from the fact that MEGA's Parsimony program has not calculated the branch lengths on the tree. Without branch lengths it cannot display the phylogram format, nor can it draw the tree as though it were rooted by the midpoint method. Instead it displays the tree as though it were rooted on the last sequence in the list. We can root the tree correctly by using the rooting tool as discussed in Chapter 7, but let's consider some other things first.

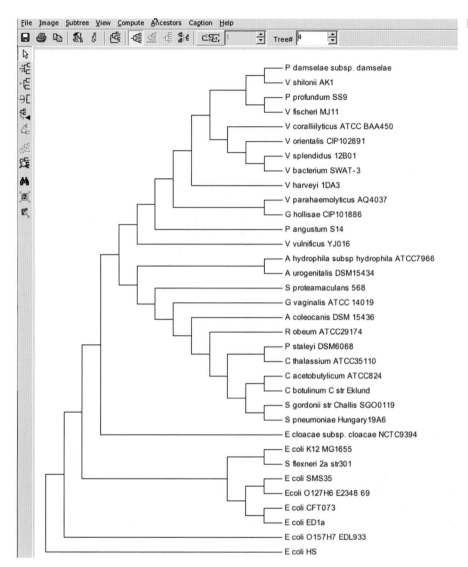

Figure 8.5

Calculating branch lengths

We mentioned earlier in the chapter that calculating branch lengths in Parsimony should be approached with caution. If it is feasible to do so, however, knowing that branch length can allow us to root the tree.

To calculate branch lengths, click the **Topology Only** button. This toggle action turns off **Topology Only** and the program will now calculate the branch lengths. Be prepared to wait awhile. For highly homoplastic sites, the number of MP paths can be almost infinite, as can be the time it takes to calculate them. Indeed, there is no way to tell in advance whether some sites in the alignment are sufficiently homoplastic to prohibit calculation of branch lengths. However, while it is calculating branch lengths MEGA displays the **Parsimony Analysis** window (**Figure 8.6**). If MEGA appears to get "hung up" while calculating branch lengths, you can click the **Stop** button to terminate the calculation and thus accept that you can only infer and show the topology of the MP tree.

Figure 8.6

Consensus and bootstrap trees

Whether we find 10 or 100 equally parsimonious trees, we are faced with the same issue: what to do with all of these trees? Clearly we can't show them all to our audience. No journal will publish even 6 nearly identical trees, and if we were to show them all to a live audience, viewers would soon fall asleep (if they were very polite). We could pick any one of the trees at random and present just that; after all, the trees are equally parsimonious, and thus equally "good." If we do this, however, we must tell our audience how many equally good, but different, trees were produced.

A better alternative is to summarize the different trees by computing a **consensus tree**. Choose **Consensus Tree** from the **Compute** menu of the **Tree Explorer** window to calculate the consensus tree. A dialog asks for a **Cut-off Value** (**Figure 8.7**). Accepting the default 50% value means that any clades occurring in fewer than 50% of the trees will not be shown, but will instead be displayed as a polytomy—multiple branches coming from one node.

Figure 8.7

| Branch | Labels | Scale | Cutoff | ◀ ▶ |

Cut-off Value for Condensed Tree 50 ▲▼ %

Cut-off Value for Consensus Tree 50 ▲▼ %

? Help   ✓ OK   ✗ Canc

Figure 8.8 shows the consensus tree of the six equally parsimonious trees found by the CNI search, correctly rooted on the Gram positive taxa. In this case there were no clades that occurred in less than 50% of the trees, so there are no polytomies.

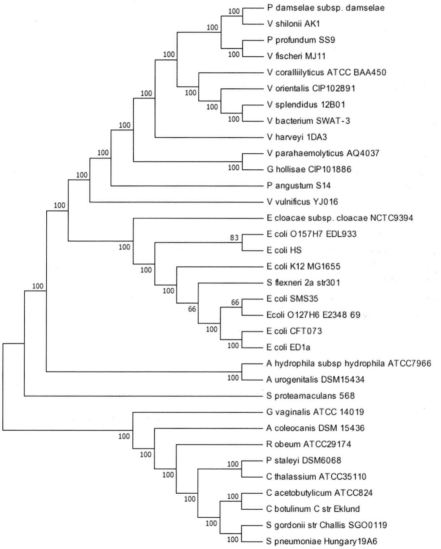

Figure 8.8

The consensus tree is not to be confused with a bootstrap tree. In this example we have looked at the consensus of six equally parsimonious trees and, in effect, averaged them. All of those trees are based on exactly the same data. Bootstrap replicates, in contrast, come from different samples of the data. **Figure 8.9** shows a bootstrap tree that was obtained by setting the **Test of Phylogeny** to **Bootstrap** in the **Analysis Preferences** window and setting the number of bootstrap replications to 500. Notice that most of the bootstrap values are very low.

Figure 8.9

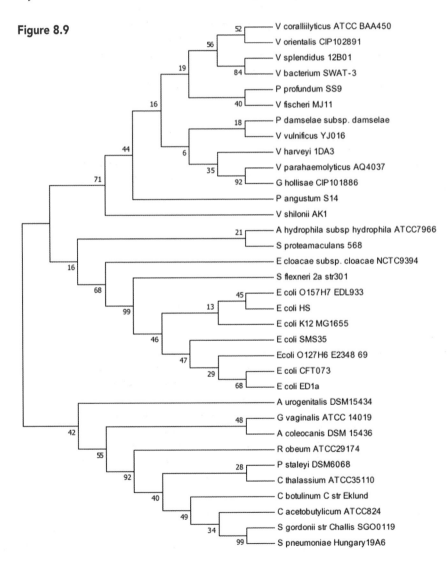

If we click the **Original Tree** tab, then click the **Compute Consensus** button, and finally click the **Bootstrap Consensus Tree** tab again, we see that the resulting Bootstrap Consensus tree (**Figure 8.10**) has very little resolution (i.e., it contains many polytomies).

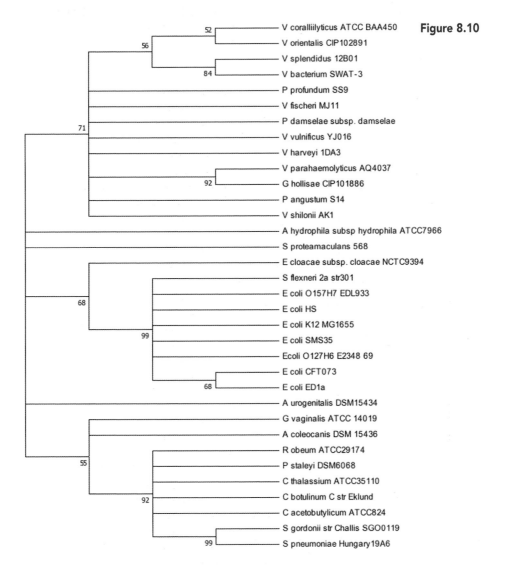

Figure 8.10

Consensus trees cannot show branch lengths, and individual trees convey no idea of how often a clade appears among the equally parsimonious trees. Which tree you choose to present is entirely up to you, but it is important to tell your audience exactly what you are showing and why.

In the Final Analysis

Does all this business of multiple trees and consensus trees mean that you should forget Parsimony and stick with NJ? Not at all. The good news about NJ is that it gives you just one tree. The bad news is that it gives you just one tree—leaving you with the impression that the tree you have is the best tree. The branching order that you see for the *E. coli/Shigella* group in the NJ tree is no more reliable than what you see in the MP tree, which is why the NJ bootstrap values are so low in that part of the tree. The number of MP trees, and any polytomies in the MP consensus tree, reflect real uncertainty about the branching order. NJ leaves you with the impression of more certainty than is really there.

Figure 8.11 shows the Bootstrap consensus tree for the LargeData set based on the CNI method. It represents the consensus of 22 trees. Compare the topology of that tree with the topology of the NJ tree in Figure 6.14.

Figure 8.11

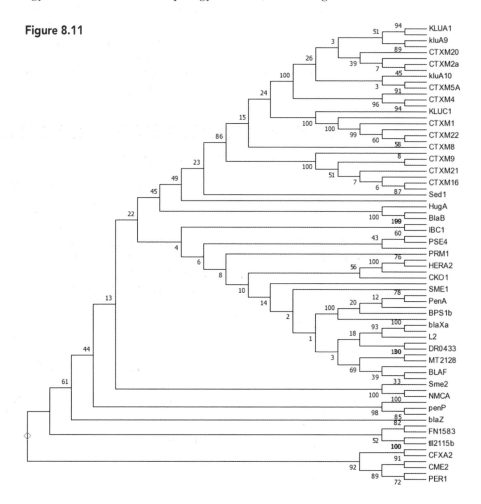

CHAPTER 9

Maximum Likelihood

Maximum Likelihood (ML) is a powerful statistical method that seeks the tree that makes the data most likely (see *Learn More about Maximum Likelihood*, below). In doing so it applies an explicit criterion—the log-likelihood of that tree—to compare the various models of nucleotide substitution for a particular data set. ML has not been used extensively outside of the fields of evolution, systematics, and phylogenetics because it has been perceived as slow and complicated. Recent advances have solved the speed problem, however, and ML is not nearly as complicated as its reputation would suggest.

LEARN MORE ABOUT

Maximum Likelihood

Maximum Likelihood (ML) tries to infer an evolutionary tree by finding that tree which maximizes the probability of observing the data. For sequences, the data are the alignment of nucleotides or amino acids. Suppose that we have four taxa and that part of the alignment of the sequences is

```
                 *
1  TCAAAAATGGCTTTATTCGCTTAATGCCGTTAACCCTTGCGGGGGCCATG
2  TCCGTGATGGATTTATTTCTGCAATGCCTGTCATCTTATTCTCAAGTATC
3  TTCGTGATGGATTTATTGCTGGTATGCCAGTCATCCTTTTCTCATCTATC
4  TTCGTGACGGGTTTATCTCGGCAATGCCGGTCATCCTATTTTCGAGTATT
```

We begin with an evolutionary model that gives the instantaneous rates at which each of the four possible nucleotides changes to each of the other three possible nucleotides (see *Learn More about Evolutionary Models*, p. 75) and a hypothetical tree of some topology and with branches of some length. For the site indicated by the asterisk (red type), there are three possible unrooted trees of four taxa (see *Learn More about Phylogenetic Trees*, p. 71), one of which looks like Figure 1.

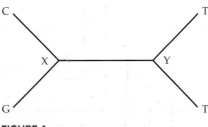

FIGURE 1

If the model being used is time-reversible, we can root the tree in Figure 1 at any node. One possible rooted tree is shown in Figure 2.

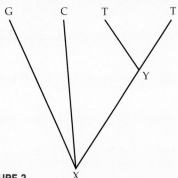

FIGURE 2

We do not know the nucleotides at nodes X and Y, but since there are four possibilities for X and four for Y, there are 16 possible scenarios that might lead to the tree in Figure 2, one of which is shown in Figure 3.

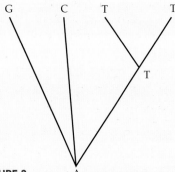

FIGURE 3

The probability of that particular scenario is the probability of observing an A at the root (P_A), which might be 1/4 or might be the overall frequency of A, depending on the model, time, and the probability of each change along the branches leading to the tips. The probability of changing from an A at the root to a G at the tip (symbolized as P_{AG}) is calculated from (1) the instantaneous rate matrix in the chosen model and (2) the length of the branch from A to G. Applying this to each branch of Figure 3, the probability of the above tree is therefore

$$P_{\mathbf{Fig3}} = P_A \times P_{AG} \times P_{AC} \times P_{AT} \times P_{TT} \times P_{TT}$$

The latter two terms are the probabilities that, given a T at the internal node T, we observe a T at the tip nodes T. They account for all possible paths which might involve multiple changes along the branch, such as T → C → T.

Another possible scenario for observing the same data is shown in Figure 4.

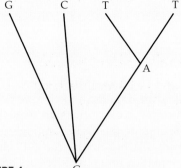

FIGURE 4

Because there are 16 such scenarios, the probabilities of each of the scenarios must be determined and added together to obtain the probability of the tree in Figure 2:

$$P_{\mathbf{Fig2}} = P_{\mathbf{Fig3}} + P_{\mathbf{Fig4}} + \dots + P_{\mathbf{Scenario16}}$$

$P_{\mathbf{Fig2}}$ is the probability for that tree for observing the data at one site, the site marked by the asterisk. The probability of observing all of the data at all of the sites is the product of the probabilities for each of the sites i from 1 to N:

$$P_{\mathrm{tree}} = \prod_{i=1}^{N} P_i$$

Because these numbers are often too small for most computers to handle, and because it is computationally easier, the probability (or likelihood) of a tree for each site i is usually expressed as a log likelihood, $\ln L_i$, and the log likelihood of the tree is the sum of the log likelihoods for each of the sites:

$$\ln L_{\mathrm{tree}} = \sum_{i=1}^{N} \ln L_i$$

The term $\ln L_{\mathrm{tree}}$ is the log likelihood of observing the alignment under the chosen evolutionary model, given that particular tree with its branching order and branch lengths. ML programs seek the tree with the largest log likelihood. In the case of four taxa, that requires calculating lnL for 15 trees, but the number of possible trees grows at a dizzying rate as the number of taxa increases. In most situations it is impossible to evaluate all possible trees, so a heuristic search method is employed to seek the most likely tree (see *Learn More about Tree-Searching*

Methods, p. 62). In practice, one usually begins by constructing a Parsimony or a Neighbor Joining tree for MEGA to use as a starting point in its search for the best ML tree.

Swofford et al. 1996 has a nice summary of Maximum Likelihood on pp. 430–431 of *Molecular Systematics.*

ML Analysis Using MEGA

As usual, Maximum Likelihood will be discussed in terms of the SmallData and LargeData data sets.

Download   Chapter 6: SmallData.meg

Download   Chapter 6: LargeData.meg

Open SmallData.meg from MEGA's main window, then choose **Construct/ Test Maximum Likelihood Tree…** from the **Phylogeny** menu to display the **Analysis Options** window (**Figure 9.1**). Many of the options are already familiar from the NJ or Parsimony **Analysis Options** window, but some are specific to ML.

Figure 9.1

Set **Test of Phylogeny, Substitutions Type,** and **Gaps/ Missing Data** as you did for NJ and Parsimony. You also need to set **Model/Method, Rates among sites, ML Heuristic Method,** and **Initial Tree for ML.** Of course, you could just accept the default options, but one of the advantages of ML is that it offers you considerable flexibility, hence the ability to set the options that are most suitable for your data. (You may wish to revisit *Learn More about Evolutionary Models,* p. 75.)

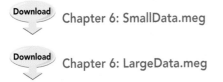

Option	Selection
Analysis	Phylogeny Reconstruction
Statistical Method	Maximum Likelihood
Phylogeny Test	
Test of Phylogeny	None
No. of Bootstrap Replications	*Not Applicable*
Substitution Model	
Substitutions Type	Nucleotide
Genetic Code Table	*Not Applicable*
Model/Method	General Time Reversible model
Rates and Patterns	
Rates among Sites	Gamma distributed with Invariant sites (G+I)
No of Discrete Gamma Categories	5
Data Subset to Use	
Gaps/Missing Data Treatment	Partial deletion
Site Coverage Cutoff (%)	95
Select Codon Positions	☑ 1st ☑ 2nd ☑ 3rd ☑ Noncoding Sites
Tree Inference Options	
ML Heuristic Method	Nearest-Neighbor-Interchange (NNI)
Initial Tree for ML	Make initial tree automatically
Initial Tree File	*Not Applicable*

✓ Compute   ✗ Cancel   ? Help

There are two options for **ML Heuristic Method**, NNI and CNI. The authors have set NNI as the default and I see no reason to dispute their judgment.

Test alternative models

MEGA offers six evolutionary models to choose from (**Figure 9.2A**) and four **Rates among Sites** (**Figure 9.2B**) for a total of 24 possible combinations.

Figure 9.2A

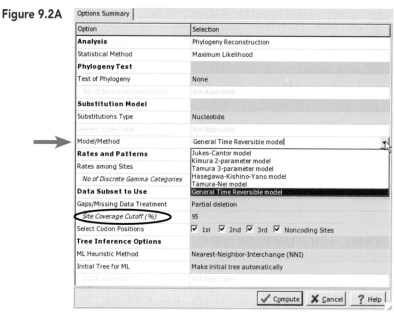

Figure 9.2B

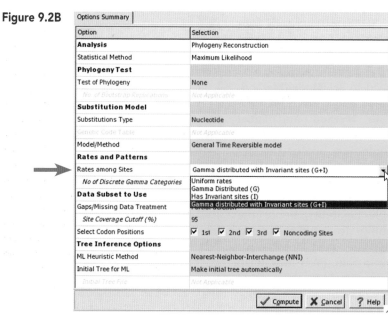

There is no practical way to know in advance which is the best combination of model and rates for your data. Unlike NJ, however, ML offers a specific way to decide which model is best: the log likelihood or, more formally, the natural logarithm of the likelihood of the data given the tree. The object is to maximize that log likelihood. You could try each of the 24 combinations to determine which gives the highest log likelihood, but MEGA provides a way to test all of those possibilities automatically.

From MEGA's main window click the **Models** menu and choose **Find Best DNA/Protein Models (ML)... (Figure 9.3)**.

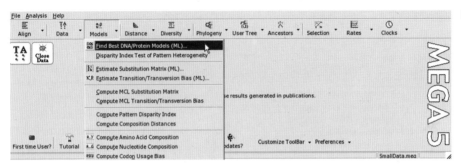

Figure 9.3

In the resulting **Analysis Preferences** window, click the **Compute** button to accept the default settings. It takes a few minutes, but eventually the **Progress** window will close and a table of the results will be displayed. **Figure 9.4** shows part of that table. A caption at the bottom of the table (not shown) explains all of the columns and abbreviations, but here we will focus on the first five columns.

Each model consists of a substitution model name plus rate description. The first model, **GTR+G**, means the General Time Reversible model with Gamma-distributed rates. The second model, **GTR+G+I** means "GTR+G

Figure 9.4

File   Edit

Table. Maximum Likelihood fits of 24 different nucleotide substitution models

Model	Parameters	AICc	BIC	lnL	(+I)	(+G)	R	f(A)	f(T)	f(C)	f(
GTR+G	74	20445.812	21004.929	-10148.515	n/a	1.03	1.16	0.297	0.249	0.205	0.2
GTR+G+I	75	20447.834	21014.495	-10148.515	0.00	1.03	1.16	0.297	0.249	0.205	0.2
K2+G	67	20512.901	21019.194	-10189.130	n/a	0.96	1.17	0.250	0.250	0.250	0.2
K2+G+I	68	20514.920	21028.761	-10189.130	0.00	0.96	1.17	0.250	0.250	0.250	0.2
TN93+G	71	20520.867	21057.347	-10189.073	n/a	0.97	1.2	0.297	0.249	0.205	0.2
TN93+G+I	72	20522.887	21066.913	-10189.073	0.00	0.97	1.2	0.297	0.249	0.205	0.2
T92+G	68	20525.144	21038.984	-10194.242	n/a	0.95	1.18	0.273	0.273	0.227	0.2
T92+G+I	69	20527.163	21048.551	-10194.242	0.00	0.95	1.18	0.273	0.273	0.227	0.2
HKY+G	70	20559.923	21088.857	-10209.612	n/a	0.96	1.19	0.297	0.249	0.205	0.2
HKY+G+I	71	20561.944	21098.423	-10209.612	0.00	0.96	1.19	0.297	0.249	0.205	0.2
JC+G	70	20713.681	21242.615	-10286.491	n/a	1.02	0.5	0.250	0.250	0.250	0.2
JC+G+I	71	20715.701	21252.181	-10286.491	0.00	1.02	0.5	0.250	0.250	0.250	0.2
GTR+I	74	21095.414	21654.531	-10473.316	0.02	n/a	0.94	0.297	0.249	0.205	0.2
GTR	73	21162.848	21714.419	-10508.043	n/a	n/a	0.93	0.297	0.249	0.205	0.2
TN93+I	71	21222.386	21758.866	-10539.833	0.02	n/a	0.95	0.297	0.249	0.205	0.2
K2+I	67	21250.912	21757.206	-10558.135	0.02	n/a	0.93	0.250	0.250	0.250	0.2
T92+I	68	21258.504	21772.344	-10560.922	0.02	n/a	0.94	0.273	0.273	0.227	0.2
HKY+I	70	21290.601	21819.535	-10574.951	0.02	n/a	0.92	0.297	0.249	0.205	0.2
TN93	70	21305.932	21834.866	-10582.616	n/a	n/a	0.94	0.297	0.249	0.205	0.2

plus Invariant Sites." The models are listed in order of best-to-worst for the data set.

The **Parameters** column shows the number of parameters that must be estimated under that model. All things being equal, it is generally considered better to estimate fewer parameters. The "things" are measures of the suitability of the model, and of course they are not always equal. Columns 3–5 show three different measures of suitability, **AICc**, **BIC**, and **lnL**. The table caption explains that models with the lowest AICc scores are considered to describe the substitution pattern the best. Likewise, the lower the BIC the better. **lnL** stands for log likelihood, and the higher the better. The table shows that for this data set GTR+G is marginally superior to GTR+G+I. Since it also has one fewer parameters, GTR+G is our best choice.

Figure 9.5

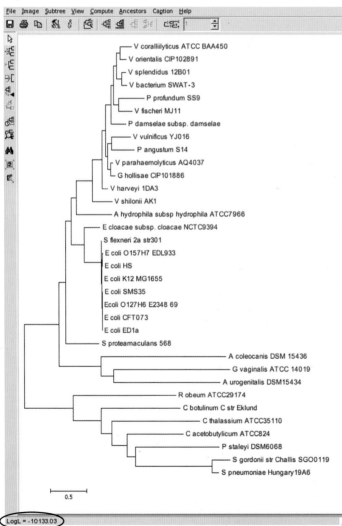

One other feature of the **Analysis Preferences** window deserves mention. The **Site Coverage Cutoff** (%) option (circled in Figure 9.2A) determines, for each site in the alignment, whether it will be considered in estimating the ML tree. If a site does not have a base in at least 95% of the sequences it is ignored. Sites that consist almost exclusively of gaps contribute little to estimating either branching order or branch length, but they do increase the computation time. The default cutoff is 95%, but you can set that to any cutoff you like.

Figure 9.5 shows the ML tree of SmallData as estimated under the GTR+G model. The log likelihood is shown at the bottom of the Tree Explorer window (circled in Figure 9.5). Notice that the actual log likelihood is slightly better than estimated in the table shown in Figure 9.4.

Figure 9.6 shows the ML tree based on the simplest model, Jukes-Cantor (JC).

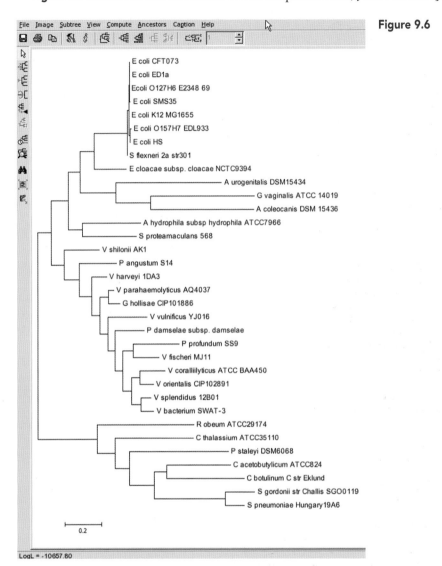

Figure 9.6

Rooting the ML tree

Recall that we rooted the NJ tree based on the Gram positive organisms (those in the lower clade on the tree in Figure 7.18) being an outgroup to the rest. The Gram positive strains formed a **monophyletic clade**; that is, they all belong to one group. In the JC tree, however the Gram positive organisms do not fall into a monophyletic clade, and thus they cannot be used as an outgroup to root the tree. In the GTR+G tree (see Figure 9.5), they do fall into a monophyletic group and can once again be used to root the tree, as is shown in **Figure 9.7**. We can thus feel assured that not only is the GTR+G tree the most likely, it is also the most consistent with our outside biological knowledge.

Compare Figure 9.7 with Figures 7.18 and 8.9.

Figure 9.7

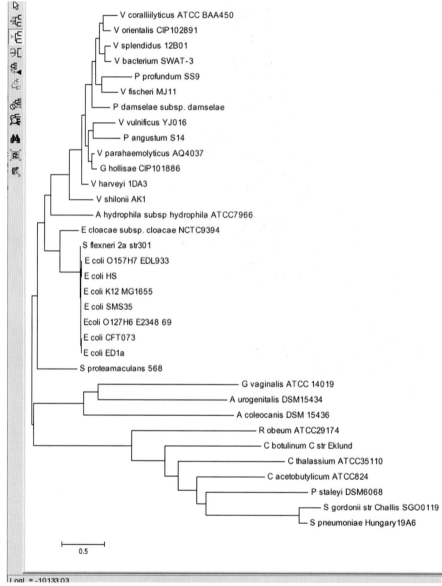

Figure 9.8 shows the rooted LargeData GTR+G+I tree and should be compared with Figures 6.14 and 8.11.

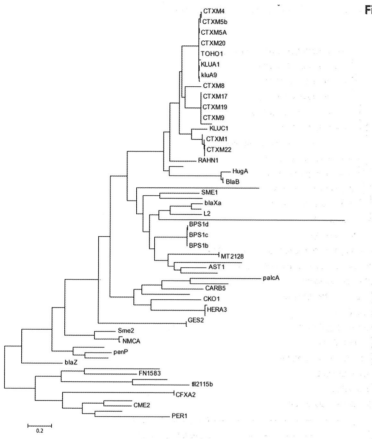

Figure 9.8

The special case of zero length branches

Figure 9.9 shows the SmallData ML tree in the cladogram format with branch lengths labeled. Notice that within the *E. coli* group there are several branches whose distances are zero. That is not an artifact of rounding; the lengths really are zero.

Figure 9.9

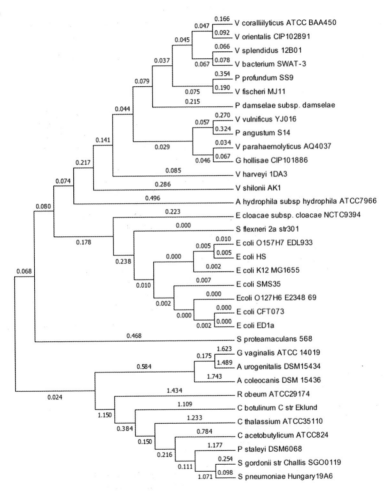

Branch lengths represent the amount of genetic change between a node and its descendant. A length of zero means that there has been no change between the node and its descendant (i.e., they are identical). With that in mind, notice that *E. coli* strains CFT073 and ED1a are identical.

When a branch length is zero, that branch should be eliminated and the descendant node moved back to its ancestral node. Phylogenetics programs are designed to estimate strictly bifurcating trees, even when branch lengths are zero, and tree-drawing programs draw the trees that way. That makes perfectly good sense when the taxa are species and it assumed that the ancestors of existing species are not actually in the sample. In this case the sample includes some individuals and their descendants. In such cases it may be important to bring those relationships to the attention of the audience.

Figure 9.10 shows the SmallData cladogram that has been redrawn manually, using a dedicated illustration program (such as Adobe Illustrator) in such a way as to make the ancestor–descendant relationships clear. Compare Figure 9.10 with Figure 9.9.

Figure 9.10

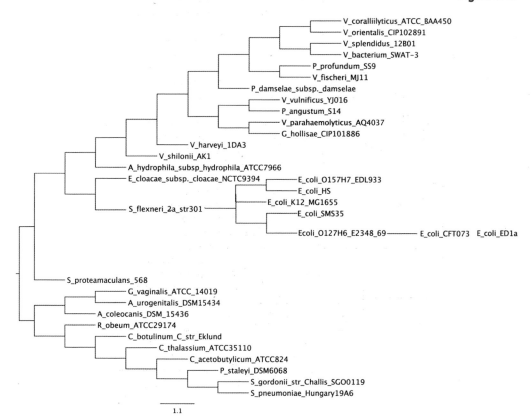

Estimating the Reliability of an ML Tree by Bootstrapping

It is just as important to know the reliability of an ML tree as it is to know the reliability of NJ or MP trees. To estimate reliability, choose **Bootstrap method** as the **Test of phylogeny** in the Analysis Preferences window (**Figure 9.11**).

Figure 9.11

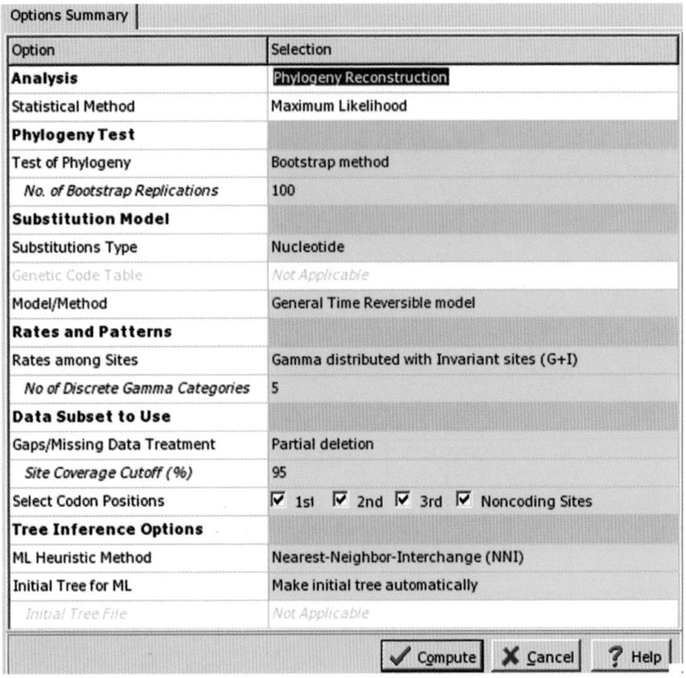

Option	Selection
Analysis	Phylogeny Reconstruction
Statistical Method	Maximum Likelihood
Phylogeny Test	
Test of Phylogeny	Bootstrap method
No. of Bootstrap Replications	100
Substitution Model	
Substitutions Type	Nucleotide
Genetic Code Table	Not Applicable
Model/Method	General Time Reversible model
Rates and Patterns	
Rates among Sites	Gamma distributed with Invariant sites (G+I)
No of Discrete Gamma Categories	5
Data Subset to Use	
Gaps/Missing Data Treatment	Partial deletion
Site Coverage Cutoff (%)	95
Select Codon Positions	☑ 1st ☑ 2nd ☑ 3rd ☑ Noncoding Sites
Tree Inference Options	
ML Heuristic Method	Nearest-Neighbor-Interchange (NNI)
Initial Tree for ML	Make initial tree automatically
Initial Tree File	Not Applicable

Options Summary

✔ Compute   ✘ Cancel   ? Help

Notice that the default number of bootstrap replicates is 100. Because it takes so much longer to estimate an ML tree than to estimate an NJ tree, I suggest you accept this default. Of course, if you have time to estimate 1000 replicates that is preferable, but 100 is certainly acceptable. **Do not set the number of replicates below 100.**

Figure 9.12 shows the SmallData strict consensus bootstrap tree, and **Figure 9.13** shows the LargeData strict consensus bootstrap tree.

Figure 9.12

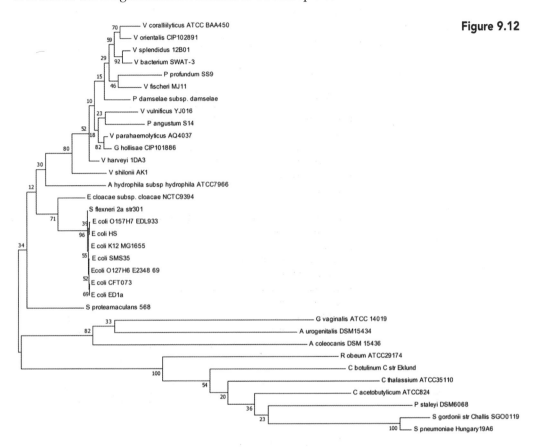

Figure 9.13

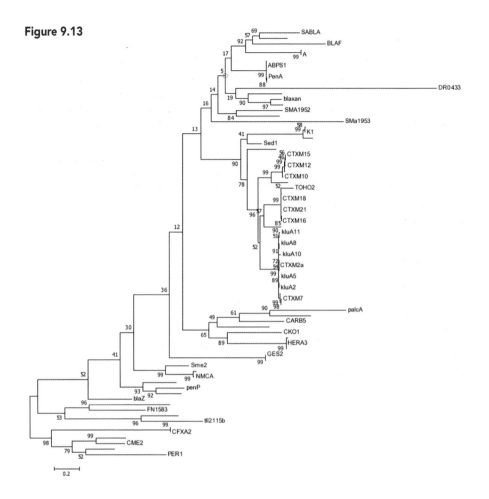

What about Protein Sequences?

To use ML with protein sequences, set **Substitutions type** to **Amino acid** in the **Analysis Preferences** window. The **Model/Method** choices change to models that are available for protein sequence analysis (**Figure 9.14**). Use the same methods as for DNA sequences to choose the best substitution model and rates (see Figure 9.3). Everything else is done as for DNA sequences.

Figure 9.14

Option	Selection
Analysis	Phylogeny Reconstruction
Statistical Method	Maximum Likelihood
Phylogeny Test	
Test of Phylogeny	None
No. of Bootstrap Replications	*Not Applicable*
Substitution Model	
Substitutions Type	Amino acid
Genetic Code Table	Standard
Model/Method	Jones-Taylor-Thornton (JTT) model
Rates and Patterns	Poisson model
Rates among Sites	Equal input model / Dayhoff model
No of Discrete Gamma Categories	Dayhoff model with Freqs. (+F) / Jones-Taylor-Thornton (JTT) model
Data Subset to Use	JTT with Freqs. (+F) model
Gaps/Missing Data Treatment	WAG model / WAG with Freqs. (+F) model
Site Coverage Cutoff (%)	95
Select Codon Positions	*Not Applicable*
Tree Inference Options	
ML Heuristic Method	Nearest-Neighbor-Interchange (NNI)
Initial Tree for ML	Make initial tree automatically
Initial Tree File	*Not Applicable*

Options Summary

Bayesian Inference of Trees Using MrBayes

Bayesian Inference (BI), like Maximum Likelihood, is a powerful and well accepted method for estimating phylogenetic trees. It differs from ML in that BI seeks the tree that is most likely given the data and the chosen substitution model, whereas ML seeks the tree that makes the data the most likely. Like ML, BI uses log likelihood as a criterion for choosing among possible trees.

Imagine a landscape in which every possible tree is represented by a point on a surface, points that are more alike are closer to each other on the surface, and each point is elevated above the plane according to its likelihood. Thus the most likely trees are near the top of some hill, and the job of a BI program is to identify those trees.

BI programs infer phylogenetic trees by (1) choosing some tree as a starting point, (2) determining that tree's likelihood, (3) changing the tree slightly by modifying the topology or slightly changing a branch length, (4) calculating the likelihood of the new tree, and (5) most of the time accepting the new tree if its likelihood is greater than that of the old tree. That five-step process constitutes a **generation** and is iterated over and over again until reaching a point where the modifications do not change the likelihoods significantly (i.e., where the likelihoods of the trees are not significantly different). At that point the program is said to have **converged** on the set of most likely trees and it will calculate and report a consensus of those trees.

MrBayes: An Overview

MrBayes is the most widely used program for estimating trees by BI.* Like any other program, it requires some means of putting in the data and some means of instructing the program exactly how to carry out its job. The input data is in the form of an alignment in Nexus format (see Appendix I). As discussed in more detail later in this chapter, an alignment can be exported from MEGA in the form of a FASTA file, then converted to the Nexus format by using the utility program

---

* Although scheduled for release in the Fall of 2010, MrBayes v3.2 was not yet available as of this book's press date in March 2011. This chapter is therefore based on a beta release of version 3.2, and some details may be different in the final version.

FastaConvert. (See Chapter 11, which describes how to install MrBayes on your computer platform and how to download and install the utility programs; you may want to read Chapter 11 before going through this chapter.)

The easiest way to enter the model and all the necessary instructions into MrBayes is in the form of a "mrbayes block" of statements. The mrbayes block is added to the Nexus file following the data (alignment) block.

Each time a tree is modified slightly and its likelihood compared with that of the previous tree is considered a generation. Every so many generations—typically 100—the tree is saved to a file and the current values of the various parameters are saved to another file. The program must know how many generations to try before calculating a consensus of the most likely trees. It has to try enough trees to reach a point where all the trees that it is currently searching among are essentially equally likely, i.e., until it has converged. If it doesn't run long enough, it won't have the chance to converge on the most likely set of trees. If it runs much longer after it has converged, it is just wasting time sampling that set of most likely trees over and over again.

Prior to Version 3.2, MrBayes required the user to specify a number of generations (`ngen`) and then to judge whether a convergence diagnostic had reached an acceptably low value. Even if the program converged long before the specified number of generations, it had to execute that number. Version 3.2 has been vastly improved by providing the option of having the run terminate whenever the program has converged—i.e., as soon as the conversion diagnostic has reached an acceptable value. In general, a value of 0.01 or less indicates convergence. Thus `ngen` now specifies a *maximum* number of generations instead of an absolute number.

MrBayes actually runs multiple **chains** of generations, each started from a different initial tree (see *Learn More about Bayesian Inference*, p. 141). The default number (`nchains`) is 4, but that can be modified by the user. Only one of those chains—the **cold chain**—really matters in the sense that it is the cold chain's tree that is saved to the trees file. So why do the other chains even exist? Every so often a chain is "swapped" and a new chain is assigned the status of being the cold chain. **Swapping** reduces the chance that the process will get stuck at a local maximum—that is, on a set of trees that are not on the highest hill. The frequency with which chains are swapped is set by the `temp` parameter; this is set to 0.2 by default, but can be changed if desired (see the section "What If You Don't Get Convergence?" on p. 152).

Once the program has converged on a set of roughly equally likely trees it must know exactly which trees to include in its consensus tree. The problem is that the program has not only saved the high-likelihood trees near the "top of the hill," it also saved the low-likelihood trees that were encountered near the beginning of the run. The program must be told how many of those early trees to discard when calculating the consensus tree. That is usually stated as a fraction of the total trees, the "burnin fraction" parameter. The default is to discard the first 25% of the saved trees (`burninfrac = 0.25`), but the user can change that fraction. Of course, the more trees that are discarded, the fewer there are to average to obtain the consensus tree.

Bayesian Inference

Bayesian inference is based on the notion of posterior probabilities—probabilities that are estimated based on some model (prior expectations) *after* learning something about the data. For instance, suppose that you have been told that 90% of the coins in a bag are true coins and 10% are biased to turn up heads 80% of the time. You are blindfolded and asked to pick a coin at random; then you are asked, "What is the probability that the coin you chose is a biased coin?" Having nothing more to go on than your model that 90% of the coins are true, your obvious answer is 0.1 based on the model you have been given.

Now suppose that you are allowed to toss the coin you chose 10 times and that you observe the following result of your tosses: HHTHHTTHHH. We will use X to symbolize that result. The probability of that result given that the coin is true—symbolized $P[X \mid \text{True}]$ where the vertical line means "given that"—is

$$P[X \mid \text{True}] = 0.5^{10} = 9.76 \times 10^{-4}$$

The probability of that result given a biased coin is

$$P[X \mid \text{Biased}] = 0.8^7 \times 0.2^3 = 1.67 \times 10^{-3}$$

The posterior probability that the coin is biased (i.e., the probability that it is biased given the result HHTHHTTHHH) is given by Bayes formula as

$$P[X \mid \text{Biased}] = \frac{P[X \mid \text{Biased}] \times P[\text{Biased}]}{P[X \mid \text{Biased}] \times P[\text{Biased}] + P[X \mid \text{True}] \times P[\text{True}]}$$

$$P[X \mid \text{Biased}] = \frac{1.67 \times 10^{-3} \times 0.1}{(1.67 \times 10^{-3} \times 0.1) + (9.76 \times 10^{-4} \times 0.9)}$$

Thus

$$P[X \mid \text{Biased}] = 0.13$$

and your estimate of the probability that this is a biased coin has increased from 0.1 to 0.13 based on your observation of results.

Bayesian analysis of phylogenies (Rannala and Yang 1996; Mau and Newton 1997; Mau et al. 1999) is similar to Maximum Likelihood in that the user postulates a model of evolution and the program searches for the best trees that are consistent with both the model and with the data (the alignment). It differs somewhat from ML in that, while ML seeks the tree that maximizes the probability of observing the data given that tree, Bayesian analysis seeks the tree that maximizes the probability of the tree given the data *and the model for evolution*. In essence, this re-scales likelihoods to true probabilities in that the sum of the probabilities over all trees is 1.0 under the Bayesian approach, which permits using ordinary probability theory to analyze the data.

Unlike ML, which seeks the single most likely tree, Bayesian analysis searches for the best set of trees. As ML searches a landscape of possible trees, it moves from point to point seeking higher points (i.e., more likely trees). If there is more than one hill on the landscape, ML can get trapped on a hill even if there is a higher hill (i.e., a better set of trees) elsewhere. While other heuristic searches do not consider the same tree more than once, the Bayesian approach will often consider the same tree many times.

MrBayes uses the Metropolis-Coupled Monte Carlo Markov Chain (MCMCMC) method, which can be visualized as a set of independent searches that occasionally exchange information. This method allows a search to leap a valley that would otherwise trap it on a suboptimal hill. The final product is a set of trees that the program has repeatedly visited; this set constitutes the top of the hill.

In principle, Bayesian inference of phylogenetic trees works the same way as the coin toss example above. In this case, the model is a tree with a specific topology (branching order) and with specified branch length, a specified stochastic model of DNA substitutions (see *Learn More about Evolutionary Models*, p. 75), and specified distribution of rates across the sites. While it is relatively easy to calculate the posterior probability of a biased coin given the frequency and expected outcomes for biased coins, it is usually not possible to calculate the

posterior probabilities of all the trees analyti-
cally. Instead, the MCMCMC method can be
used to sample trees from the distribution of
posterior probabilities.

Like Parsimony and Maximum Likelihood,
the Bayesian method is character-based and
is applied to each site along the alignment. It
begins with a tree (either a user-specified tree
or a randomly chosen tree) with a combination
of branch lengths, substitution parameters, and
rate variation across sites parameter to define
the initial state of a chain. A new state of the
chain is then proposed and the probability of
the new state, given the old state, is calculated.
A random number between 0 and 1 is drawn,
and if that number is less than the calculated
probability, the new state (new tree) is accepted;
otherwise the state remains the same. This constitutes a single generation of the chain.

The proposed new state involves moving a
branch and/or changing the length of a branch
to create a modified tree. If the new tree is more
likely than the existing tree, given the model
and the data, it is more likely to be accepted.
Notice, however, that while the general trend
will be toward accepting increasingly more likely
trees, not every move will pick a more likely
tree, and some moves will pick a less likely tree.

As the number of generations increases, the
process closes in on a set of trees in which the
likelihoods are so similar that accepting or reject-
ing a change is essentially a random choice. At
this point the chain has converged on a *stable
likelihood value*. Just as fair and biased coins
were not equally frequent in the example above,
the different trees are not equally frequent in
the distribution of choices after the chain has
converged. Given a large enough number of
samples, the frequency with which the various
choices are sampled is almost exactly the frequency of those trees in the likelihood distribution, and the frequency with which any particular
tree is sampled is just about the probability that
it is the best tree among the roughly equally
likely trees. The sampled trees are saved, and
the frequency of any particular tree in that set of
saved trees is taken as the probability that it is
the best of the equally likely trees.

It is possible for Maximum Likelihood to get
fixed at the top of a hill that is not the highest
hill (i.e., a hill that does not contain the most
probable tree); the same thing can happen
with the Bayesian approach. MrBayes deals
with that possibility by running several (typically four) *independent* chains starting from
different initial trees. Because of the stochastic
element involved in deciding whether or not to
accept a proposed new state, the chains quickly
diverge from one another. One of the chains,
designated the "cold chain," is the source of
the tree that is saved once every so many (typically 100) generations. Each generation there
is a probability-based chance for two chains to
swap states. This means that if the cold chain
has gotten stuck at the top of a low probability
hill, it has a chance to escape by swapping with
another chain that may be on a higher hill. This
does not insure that the set of saved trees will
represent those near the top of the highest
probability hill, but it helps a lot. ◄€⋮

Saving time (and perhaps your sanity)

BI is a computationally intensive method. The time required can be hard to
predict; I have run analyses that took a few minutes and I've run analyses that
took several weeks. Anything that speeds up the process is helpful. Usually
the program is initiated with a random tree that is unlikely to resemble the
final trees very much. Considerable time is therefore spent bumbling around
before getting near the most likely set of trees. What if, instead of starting with
a random tree, you could start with a pretty good tree? A lot of bumbling time
can be saved by employing a **user tree**. The user tree is estimated by some

other, faster method, such as NJ or ML, exported in the Newick format (see Appendix I), and added to the Nexus alignment file in the form of a "trees block" just after the data block and before the mrbayes block. Not only does a user tree speed up the process, in some cases I have found that without a starting user tree I cannot obtain convergence.

Even starting with a user tree, some data sets, particularly those involving hundreds of taxa or very long sequences, can take several days or longer to run. Fortunately, MrBayes provides a continuous estimate of the time remaining to reach the maximum number of generations. Imagine your distress when, just as you are approaching that point, your computer crashes or the power goes out. Prior to version 3.2 your only option was to start over while venting your most virulent expletives. Version 3.2 has an option that in effect allows you to save the status of everything every few thousand generations. You can then restart as of that last save, and typically you wind up losing an hour instead of several days.

Choose a model

Like Maximum Likelihood, Bayesian Inference requires that you specify a model of nucleotide substitutions (see *Learn More about Evolutionary Models*, p. 75). Use MEGA's model selection function, described in Chapter 9, to choose the model that is best suited to your data. **Table 10.1A** shows the equivalency between model specification in MEGA and in MrBayes as specified in the `lset` command.

TABLE 10.1   Equivalency between MEGA and MrBayes

(A) MODEL		(B) RATES	
MEGA	MrBAYES	MEGA	MrBAYES
JC	nst=1	Uniform	rates = equal
K2, HKY, T3P, TN93	nst=2	Gamma Distributed (G)	rates = gamma
GTR	nst=6	Has Invariant Sites (I)	rates = propinv
		Gamma Distributed with Invariant Sites (G+I)	rates = invgamma

A General Strategy for Estimating Trees Using MrBayes

The above considerations suggest a general strategy for efficiently estimating trees with MrBayes v3.2.

1. Align a set of sequences in MEGA and save the alignment in the `.mas` format.

2. From the **Data** menu of the **Alignment Explorer** choose **Export Alignment**. From the submenu choose **FASTA format** and save the file. Use the Utility program FastaConvert (see Chapter 11) to convert the alignment in FASTA format to an alignment in the Nexus format that can be read by MrBayes.

FastaConvert will save the alignment with the extension `.nxs`. (MEGA can export the alignment in the Nexus format, but only the "PAUP 3.0" version of that format will work with MrBayes, and even that file would require additional editing.)

3. From the **Data** menu of the **Alignment Explorer** again choose **Export Alignment**. From the submenu choose **MEGA format**, then open the resulting `.meg` file from the **File** menu of MEGA's main window.

4. In MEGA's main window choose **Find Best DNA/Protein Models (ML)** to identify the model best suited to your data. Estimate an ML tree using that model as described in Chapter 9. Do *not* estimate a bootstrap tree.

5. When the **Tree Explorer** window opens, choose **Export Current Tree (Newick)** from the **File** menu and save the file with an appropriate name. This will be the user tree for MrBayes.

6. In your favorite text editor add a "trees block" and a "mrbayes block" to the `.nxs` file and save it with the extension `.bay` to remind you that it is an input file for MrBayes.

7. Execute the `.bay` file

There are some additional details to consider, but that covers the general approach.

 Chapter 10: MrBayes Templates

MrBayes is a command line program and does not use pulldown menus to enter the various commands. Instead it depends completely on what you type in the mrbayes block. That makes it very vulnerable to typographical errors, forgetting an essential semicolon, etc. To minimize the opportunity for such frustrating errors, I have provided a file of template mrbayes blocks that are suitable to most of the situations you are likely to encounter. You can copy one of those blocks into the `.nxs` file, then modify the block to suit your situation and create the `.bay` execution file. Included among those is a template "trees block" into which you can paste the Newick format ML tree you estimated. (See Appendix I to learn about blocks in the Nexus format.)

Creating the Execution File

The execution file is simply an alignment in Nexus format with a "block" of MrBayes commands added after the data block. To modify the Nexus file for MrBayes, use any text editing program to open the .nxs file; then type or paste a mrbayes block into the .nxs file following the `end;` statement at the end of the data block. If you use a user tree, you must add a trees block between the data block and the mrbayes block. After adding the mrbayes block, I often rename the file with a `.bay` extension.

A mrbayes block begins with `begin mrbayes;` and ends with `end;`. The semicolon at the end of each of those elements is essential. Between the `begin` mrbayes; and end; are a series of statements, each ending with a semicolon, that tell MrBayes what to do with the data. Each statement consists of a **command** and one or more **options**, and is terminated by a semicolon. A statement may extend over more than one line. A mrbayes block thus looks like this

```
begin mrbayes;
 command option;
 command options;
end;
```

The simplest block is one suited for non-coding sequences and not employing a user tree. Let's begin with that and then consider more complicated blocks.

```
begin mrbayes;
 log start replace;
 set autoclose = no nowarn=no;
 lset applyto = (all) nst=6 rates = invgamma;
 prset applyto = (all);
    [For the JC model add to the prset command the option
     statefreqpr = fixed(equal]
 unlink revmat=(all) shape=(all) pinvar=(all)
    statefreq=(all) tratio= (all);
 showmodel;
 mcmc ngen=10000000 printfreq=1000 samplefreq=100
    nchains=4 temp=0.2 checkfreq = 50000 diagnfreq = 1000
    stopval = 0.01 stoprule = yes;
 sumt relburnin = yes burninfrac = 0.25
    contype = halfcompat;
 sump relburnin = yes burninfrac =0.25;
 log stop;
end;
```

 Chapter 10: MrBayes example block

It will be helpful for you to open the MrBayes example block file with your favorite text editor program (*not* Word or WordPerfect; see Chapter 1) and to look at on screen as I discuss each statement in detail below.

What the statements in the example mrbayes block do

The log start replace; statement starts recording everything that appears in the MrBayes window to a file. Since no file name was specified, the file will have the default name log.out. If we wanted a different name, such as myfile.log, we could have included the option filename = myfile.log before the semicolon. The reason for recording the log will be discussed later in this chapter.

In the statement `set autoclose = no nowarn=no;` the option "autoclose = no" tells MrBayes that if it has reached the maximum number of generations, it should ask if you want to run more generations and wait for an answer before continuing. The "nowarn= no" option means that MrBayes will remind you if any existing files of the same name will be overwritten. If you prefer not to be reminded, set `nowarn = yes`.

The statement `lset applyto = (all) nst=6 rates = invgamma;` sets the parameters for the model you use. The option `applyto = (all)` ensures that, if the data are partitioned, the settings will apply to all partitions. The option `nst=6` sets the number of states (`nst`) to 6, which is the GTR (General Time Reversible) model (see Table 10.1A); `rates = invgamma` sets across-site rate variation to the gamma distribution with a proportion of invariant sites (see Table 10.1B). By default MrBayes estimates the gamma-shape parameter that determines the proportion of invariant sites.

The command `prset applyto = (all);` specifies that the default priors will apply to all partitions. In this case there is (by default) a single partition.

MrBayes treats anything within square brackets as a comment and ignores it, so `[For the JC model add to the prset command the option statefreqpr = fixed(equal]` is a reminder that if you want to use the JC model (`nst = 1`) you need to add the `statefreqpr` option to the `prset` command.

The command `unlink` ensures that the various parameters are estimated independently for each of the parameters. Thus, `unlink revmat=(all) shape=(all) pinvar=(all) statefreq=(all) tratio=(all);` lists the parameters that should be independently estimated.

The `showmodel` command causes all of the model settings, including the default settings that we have not specified, to be printed to the screen and to the log file. This can be useful later on when you are trying to remember exactly how you did this particular analysis. This is also an example of a command that requires no options.

The command `mcmc` tells MrBayes how to run the analysis. When the mcmc statement is executed, the run actually begins. The `mcmc` command has no fewer than 37 different available options; the statement doesn't end until the `stoprule = yes` option. Most options have default settings that are just fine for most situations. You can see a complete list of all 37 options and a discussion of each by starting MrBayes and typing `help mcmc`.

The `mcmc` options shown in the example block are:

- The `ngen=10000000` option says to run the analysis for a maximum of 10,000,000 generations

- The `printfreq = 1000` option says to print the results to the screen every 1000 generations. Printing more frequently wastes time and it makes the log file huge.

- The `samplefreq = 100` option says to record the current tree and parameter values to files every 100 generations.

- The `nchains = 4` option says run four independent chains. There is little to be gained from running more chains, and more chains increases the running time.

- The `temp = 0.2` option sets the temperature for heating the chains (i.e., the frequency with which chains are swapped) to 0.2.

- The `checkfreq = 50000` option says to save the status of the analysis every 50000 generations just in case the computer crashes or the power fails. Later I'll describe how to recover from such crashes.

- The `diagnfreq = 1000` option says to calculate the convergence diagnostic every 1000 generations.

- The `stopval = 0.01` option says that convergence has been reached when the value of the convergence diagnostic is ≤0.01.

- The `stoprule = yes` option says to stop when the `stopval` has been reached. When the stoprule is in effect, the run may end before the maximum number of generations (`ngen`) is reached.

Taken together, the last three options on the above list save time by preventing the program from running useless generations after convergence has occurred.

The `sumt` command tells MrBayes how to summarize the saved trees—in other words, how to write the consensus tree after convergence has been reached. The early trees (those with low likelihoods) need to be discarded. Experience has shown that it is sufficient to discard the first 25% of those trees obtained before convergence occurred. Since we don't know in advance how many generations will be required for convergence, we use the `relburnin = yes` option to determine the number to be discarded as a proportion of the trees, not as an absolute number. The `burninfrac = 0.25` option says to discard the first 25% of the trees. The `contype = halfcompat` option tells MrBayes what kind of consensus tree to write: `halfcompat` means a majority rule consensus tree, and the alternative `allcompat` means a strict consensus tree. Recall that a strict consensus tree is strictly bifurcating, whereas a majority rule consensus collapses branches whose posterior probabilities (see *Learn More about Bayesian Inference*, p. 141) are less than 0.5 to polytomies. The consensus tree is written to a file with the extension `.con`.

The `sump` command specifies how to summarize the parameter values that were saved when the trees were saved. The `relburnin = yes` and `burninfrac = 0.25` options serve exactly the same purpose as they do in the sumt command. The output from `sump` is printed to the screen and, because of the `log start` statement, is saved to the `log.out` file. In some circumstances it can be useful to save the `sump` output to a separate file by including the `printtofile = yes` option within the `sump` statement.

The `log stop` command just says to quit recording everything to the log file.

How the stoprule option of the mcmc command is implemented

The `nruns` option in the mcmc command determines the number of simultaneous runs that MrBayes performs. The default value is 2, which means that MrBayes normally runs two completely independent runs, each run consisting of four chains. Because of that default value it is not necessary to explicitly specify `nruns` in the mcmc command. The convergence diagnostic is obtained by comparing the two runs. Each run saves its trees and parameters in separate files, so when `sumt` and `sump` summarize the trees and parameters, the results of the two runs are summarized together. If at convergence there have been 400,000 generations and the burnin fraction was 0.25, then 3000 trees from each run will be summarized to estimate the consensus tree. Why 3000, not 300,000 trees? Because the `samplefreq` (sample frequency) was set to 100—that is, a tree was saved every 100 generations.

How Do You Run a MrBayes Analysis?

So far everything in this chapter has been like reading a description of driving a car without being told how to start the engine. Let's deal with that, then come back to more complex mrbayes blocks.

The first step is to install MrBayes, as described in Chapter 11.

MrBayes is run from within the Command Prompt program (Windows) or the Terminal program (Macintosh and Linux). See Chapter 11 for a discussion of using and navigating within those programs. In either Terminal or Command Prompt, navigate to the folder that contains the input execution file. The "execution file" or ".bay file" is the Nexus alignment with the mrbayes block. Type `mb` to start MrBayes. Assuming that the execution file is named `myfile.bay`, at the MrBayes prompt (`>`) enter `exe myfile.bay`. That's it. MrBayes starts executing the commands in the mrbayes block. With the `stoprule` in effect the run will continue until the maximum number of generations is reached or convergence has occurred. If convergence occurs, the `sumt` and `sump` statements are executed, files are written, and the run is complete.

What if convergence has not occurred after `ngen` generations? With `autoclose = no` in effect MrBayes will ask if you want to end the run now or if you want to run for more generations. If you decide to end the run, the program moves on to the `sumt` and `sump` statements despite the fact that it has not converged. If you decide to do more generations, it asks how many and then continues as before. (These options are discussed in more detail later in the chapter.)

While MrBayes is running, it prints information to the screen (and records it to the log file). You can monitor the progress of the run from that information (see Figure 10.2, p. 151).

More Complex (and More Useful) MrBayes Blocks

The example block was chosen for its simplicity in order to explain the major commands. Now let's consider blocks that (1) include a user tree and (2) most effectively estimate trees from coding sequences.

Including a user tree

Two modifications are required to include a user tree: (1) the addition of the tree in the form of a **trees block**; and (2) the addition of a `startvals` statement to the mrbayes block. A user tree *must* be strictly bifurcating; you can't use a consensus tree that has polytomies as your user tree.

THE TREES BLOCK The trees block goes after the data block and before the mrbayes block and looks like this:

```
begin trees;
[This tree is a ML tree from the same alignment]
 tree user =(E_coliK12_IMDH:1.75440400,(((X_campestris_
NAD:0.3…;
end;
```

Like the data block and the mrbayes block, the tree block begins with `begin` and ends with `end;`. The second line is a comment, enclosed in square brackets, to remind me of the origin of the tree. That line is not required.

The next line must begin with the word "tree." The next word is the name of the tree; I've chosen to name it "user." Following that name is an equal sign, then the tree description itself in Newick format. I have indicated most of the Newick tree description by an ellipsis (…). The tree description *must* end with a semicolon. The tree description should include branch lengths, otherwise MrBayes must use random lengths to start out. The description *may not* include node labels such as clade confidences.

THE STARTVALS COMMAND AND ISSUES WITH BRANCH LENGTHS The `start-vals` command in the mrbayes block tells MrBayes to set the initial tree topology and branch lengths to those of the user tree. The command, which must appear before the `mcmc` command, is `startvals tau = user V = user;` where `tau` is the topology, `V` is the branch lengths, and `user` is the name of the tree in the trees block. If the user tree does not include branch lengths, omit `V = user` from the command.

Branch lengths in user trees sometimes cause problems. In particular, if any branches have a length of zero an error will occur and MrBayes will terminate. Sometimes the error warns you that the `"User tree 'user' does not have branch lengths so it cannot be used in setting parameter 'V{all}'"`. Other times MrBayes will freeze up as it attempts to divide by zero. In either case, the solution is to change all zero length branches to something else. Simply replace all occurrences of 0.00000000 with 0.00000001.

The nperts option of the mcmc command

The user tree starts all four chains of both runs from the same point. Since the idea is to explore all of the roughly equally probable "most likely" trees, if the starting tree is really good it can result in a failure to fully explore the relevant tree space. If you have used the `stoprule` option (see p. 147), you may well find that the first convergence diagnostic falls below the `stopval` and the run terminates too quickly. The remedy is to slightly perturb the starting tree in each chain of both runs by setting the `nperts` ("number of perturbations") option of the `mcmc` command.

Typically, during the initial part of the run the convergence diagnostic will move around, increasing and decreasing slightly before starting to decrease slowly toward the `stopval` cutoff. Usually if the run goes past 20,000 generations it will not converge too quickly and will fully explore the treespace. If it does converge too quickly, I usually set `nperts` = 2 and, if that doesn't work, I increase it to 3 or 4.

Coding sequences and the charset statement

It is useful to *partition* the characters in the alignment as being first, second, or third position in a codon and to allow MrBayes to estimate different substitution rates for these positions. Doing so can improve the resolution of the tree. We know that third positions usually evolve fastest because many third-position changes are silent, while second positions usually evolve most slowly because all second-position changes result in amino acid substitutions (many of which are eliminated by purifying selection).

To partition the characters we add the following statements to the mrbayes block before the `lset` statement:

```
charset 1st_pos = 1-.\3;
charset 2nd_pos = 2-.\3;
charset 3rd_pos = 3-.\3;
partition by_codon = 3:1st_pos,2nd_pos,3rd_pos;
set partition = by_codon;
```

A `charset` statement defines the name of a character set and assigns a subset of the characters to that character set. The statement `charset 1st_pos = 1-.\3;` defines a character set whose name is `1st_pos` and whose members are every third character starting at character 1 in the alignment—i.e., the first position in each codon. The other charset commands likewise define character sets that are the second and third characters in codons.

The `partition by_codon = 3:1st_pos,2nd_pos,3rd_pos;` statement defines a partition whose name is `by_codon`, whose membership includes the three character sets that are named `1st_pos`, `2nd_pos`, and `3rd_pos` respectively.

The `set partition = by_codon` statement actually puts the partition parameters into effect.

The priors have already been applied to all partitions by the `prset applyto = (all)` statement.

 Chapter 10: SmallData.bay

SmallData.bay is an example of a coding sequence data set that uses an ML tree as the user tree to start the MrBayes analysis. **Figure 10.1** shows the mrbayes block of SmallData.bay.

Figure 10.1

```
begin mrbayes;
    log start replace;
    set autoclose = no nowarn=no;
    charset 1st_pos = 1-.\3;
    charset 2nd_pos = 2-.\3;
    charset 3rd_pos = 3-.\3;
    partition by_codon = 3:1st_pos,2nd_pos,3rd_pos;
    set partition = by_codon;
    lset applyto = (all) nst=6 rates = gamma;
    prset applyto = (all); [For JC add option state freqpr = fixed(equal)
    unlink revmat=(all) shape=(all) pinvar=(all) statefreq=(all) tratio= (all);
    showmodel;
    startvals tau = user V = user;
    mcmc ngen=10000000 printfreq=1000 samplefreq=100 nchains=4 temp=0.2 checkfreq = 50000 diagnfreq = 1000 stopval = 0.01 stoprule = yes nperts = 1;
    sumt relburnin = yes burninfrac = 0.25 contype = halfcompat;
    sump relburnin = yes burninfrac =0.25;
    log stop;
end;
```

The Screen Output while MrBayes Is Running

To estimate a BI tree from the SmallData.bay data set, in Terminal or Command prompt navigate to the folder that contains the execution (`.bay`) file, enter `mb`, and when MrBayes starts enter `exe SmallData.bay`. MrBayes immediately prints a lot of stuff to the screen, most of which is a description of the data and the conditions under which the tree will be estimated. It then begins to print the chain results (**Figure 10.2**). Those results allow you to monitor the progress of the run. Because the `printfreq` option of the `mcmc` command is set to 1000, it only prints the results to the screen once every 1000 generations.

Figure 10.2

```
Chain results:
    1 -- [-13042.419] (-12952.512) (-12949.936) (-13065.567) * [-12950.768] (-12977.346) (-12961.093) (-12952.144)
 1000 -- (-11625.253) (-11675.595) [-11618.824] (-11800.745) * [-11687.224] (-11758.636) (-11743.655) (-11844.404) -- 1:23:15

Average standard deviation of split frequencies: 0.137894

 2000 -- (-11168.832) (-11253.225) (-11242.923) [-11196.881] * (-11239.869) (-11366.187) (-11323.839) (-11233.989) -- 1:23:10

Average standard deviation of split frequencies: 0.089995

 3000 -- (-11034.395) (-11041.159) (-11097.853) [-11029.857] * (-11036.977) (-11100.200) (-11098.671) (-11040.639) -- 1:23:05

Average standard deviation of split frequencies: 0.102383

 4000 -- (-10973.192) (-10967.929) (-11016.952) [-10950.345] * [-10946.504] (-10964.806) (-11004.372) (-10978.191) -- 1:23:00

Average standard deviation of split frequencies: 0.122975

 5000 -- (-10924.771) (-10932.028) (-10960.445) [-10890.201] * [-10880.156] (-10911.985) (-10963.246) (-10911.388) -- 1:22:55

Average standard deviation of split frequencies: 0.105846

 6000 -- (-10903.271) (-10900.537) (-10910.414) [-10875.393] * (-10866.048) (-10907.735) (-10928.081) [-10884.119] -- 1:22:50

Average standard deviation of split frequencies: 0.109001
```

Each result line shows the generation number at the left, then a series of four numbers each enclosed in brackets. Those numbers are the log likelihoods of the trees in each of the four chains in run 1. The cold chain is enclosed in square brackets. After the * symbol there are another four numbers showing the log likelihoods of the chains in run 2. Finally, following "—" is an estimate of the time remaining that will be required to reach the maximum number of generations specified by `ngen`. You should see the log likelihoods increasing as the number of generations increases. Notice that the log likelihoods are all negative values, so –10911 is larger than –12977.

Because the `diagnfreq` was also set to 1000, the convergence diagnostic is also printed once every 1000 generations on the line below the log likelihoods. The convergence diagnostic is the average standard deviation of split frequencies. That value generally should decrease with increasing generations, but you should not worry if it fluctuates a bit, sometimes increasing before it starts to fall again.

What If You Don't Get Convergence?

If the `stopval` for the convergence diagnostic has not been reached after `ngen` generations, then convergence has not been obtained and MrBayes will ask if you want to add more generations (red arrow in **Figure 10.3**).

Figure 10.3

```
995000 -- [-10845.502] (-10858.122) (-10841.802) (-10869.266) * (-10869.861) (-10881.610) (-10855.285) [-10849.682] -- 0:00:25
Average standard deviation of split frequencies: 0.031043
996000 -- [-10842.519] (-10859.043) (-10846.925) (-10867.437) * (-10849.296) (-10885.027) (-10906.879) [-10847.452] -- 0:00:20
Average standard deviation of split frequencies: 0.031003
997000 -- [-10841.095] (-10868.903) (-10863.278) (-10862.174) * (-10857.512) (-10848.017) (-10889.632) [-10856.390] -- 0:00:15
Average standard deviation of split frequencies: 0.030979
998000 -- (-10857.278) (-10877.416) (-10850.806) [-10848.611] * [-10836.727] (-10856.154) (-10870.159) (-10885.436) -- 0:00:10
Average standard deviation of split frequencies: 0.030918
999000 -- (-10837.620) (-10869.543) (-10851.866) [-10836.812] * [-10853.954] (-10850.829) (-10871.434) (-10873.795) -- 0:00:05
Average standard deviation of split frequencies: 0.030925
1000000 -- [-10838.937] (-10855.540) (-10862.531) (-10837.460) * (-10845.218) (-10849.458) [-10820.863] (-10854.832) -- 0:00:00
Average standard deviation of split frequencies: 0.030809

Continue with analysis? (yes/no): y
    Additional number of generations: █  ◄───────
```

The easiest choice is to add more generations but, depending on how long the analysis has required up to this point, this may or may not be practical. If you started with 1,000,000 generations and that required 10 hours, you may be willing to try 3,000,000 generations. But what if at that point you still don't have convergence? If the convergence diagnostic seems to have been stable for many generations and it is still >0.01, it is unlikely that more generations will help. In that case choose not to continue, let the run complete, and consider how you might modify conditions to obtain convergence on your next try. The `log.out` file includes several tables that can help you figure out whether convergence occurred and what you might do about it. Open that file in your favorite text editor.

The first thing to do is to scroll down until you see the `Chain swap in-formation` (**Figure 10.4**). In the upper potion of each matrix, just above the diagonal (circled in Figure 10.4), most of the proportions of successful state exchanges should be in the range 0.1–0.7. If those values are too low, you can decrease the `temp` setting in the `mcmc` command. I usually try `temp = 0.1`, and if that fails, `temp = 0.05`. The values in Figure 10.4 are all within the acceptable range. If those acceptance rates are outside the optimal range, it does not affect the validity of the analysis; it just takes longer to reach convergence.

```
Chain swap information for run 1:
```
Figure 10.4

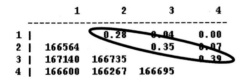

```
              1       2       3       4
      ---------------------------------
    1 |          0.28    0.04    0.00
    2 |  166564          0.35    0.07
    3 |  167140  166735          0.39
    4 |  166600  166267  166695
```

```
Chain swap information for run 2:

              1       2       3       4
      ---------------------------------
    1 |          0.27    0.04    0.00
    2 |  166789          0.33    0.07
    3 |  166482  166525          0.39
    4 |  166502  166824  166879
```

```
Upper diagonal: Proportion of successful state exchanges between chains
Lower diagonal: Number of attempted state exchanges between chains
```

The second thing to do is to find the table that summarizes the samples of parameter values (**Figure 10.5**) and look at the last column, labeled PSRF. The values in that column should be between 1.0 and 1.02. If all the values are within that range and the convergence diagnostic was pretty close to, but not quite at, 0.01, it is probably safe to use the analysis. Values outside that range confirm the failure to obtain convergence and indicate that you should try again with different settings.

Figure 10.5

Model parameter summaries over the runs sampled in files
"SmallData.bay.run1.p" and "SmallData.bay.run2.p":
(Summaries are based on a total of 15004 samples from 2 runs)
(Each run produced 10002 samples of which 7502 samples were included)

Parameter	Mean	Variance	95% Cred. Interval Lower	95% Cred. Interval Upper	Median	PSRF *
TL{all}	15.037493	0.795568	13.318010	16.815930	15.009650	1.002
r(A<->C){1}	0.327095	0.001060	0.264292	0.392198	0.327126	1.000
r(A<->G){1}	0.200140	0.000557	0.157021	0.247755	0.199137	1.001
r(A<->T){1}	0.064192	0.000395	0.029683	0.106819	0.062503	1.000
r(C<->G){1}	0.079961	0.000266	0.049218	0.114002	0.079082	1.000
r(C<->T){1}	0.249708	0.000945	0.191972	0.310798	0.248851	1.000
r(G<->T){1}	0.078904	0.000261	0.049827	0.113987	0.077932	1.000
r(A<->C){2}	0.194084	0.000985	0.137759	0.257296	0.193324	1.000
r(A<->G){2}	0.203335	0.000985	0.146269	0.266521	0.202415	1.005
r(A<->T){2}	0.036299	0.000104	0.018984	0.057748	0.035410	1.000
r(C<->G){2}	0.331347	0.002030	0.245779	0.418081	0.331109	1.001
r(C<->T){2}	0.156182	0.000748	0.106109	0.214466	0.153850	1.003
r(G<->T){2}	0.078753	0.000438	0.042246	0.124078	0.076881	1.001
r(A<->C){3}	0.095742	0.000254	0.066405	0.128263	0.095110	1.000
r(A<->G){3}	0.249717	0.000588	0.204266	0.298721	0.249057	1.002
r(A<->T){3}	0.110111	0.000231	0.081092	0.141078	0.109290	1.000
r(C<->G){3}	0.112570	0.000286	0.080503	0.146716	0.111982	1.000
r(C<->T){3}	0.365850	0.000703	0.314623	0.419533	0.365542	1.000
r(G<->T){3}	0.066010	0.000184	0.040718	0.094294	0.065297	1.000
pi(A){1}	0.242012	0.000180	0.216496	0.269096	0.241810	1.000
pi(C){1}	0.211635	0.000189	0.186870	0.240003	0.211038	1.000
pi(G){1}	0.359498	0.000339	0.323796	0.396230	0.359345	1.001
pi(T){1}	0.186856	0.000236	0.158756	0.218278	0.186427	1.000
pi(A){2}	0.403308	0.000432	0.361912	0.444086	0.403189	1.000
pi(C){2}	0.187078	0.000196	0.161244	0.215857	0.186549	1.000
pi(G){2}	0.147158	0.000171	0.123477	0.174082	0.146542	1.001
pi(T){2}	0.262456	0.000507	0.220170	0.308379	0.261812	1.000
pi(A){3}	0.234487	0.000146	0.211439	0.259673	0.234195	1.000
pi(C){3}	0.230467	0.000115	0.210005	0.252070	0.230501	1.000
pi(G){3}	0.241041	0.000172	0.216446	0.266609	0.240708	1.001
pi(T){3}	0.294005	0.000178	0.268058	0.321016	0.294075	1.000
alpha{1}	1.430900	0.033772	1.108307	1.819803	1.416271	1.000
alpha{2}	1.000148	0.013748	0.785352	1.245481	0.995254	1.000
alpha{3}	1.345286	0.108895	0.854110	2.134668	1.291055	1.003

* Convergence diagnostic (PSRF = Potential scale reduction factor [Gelman
and Rubin, 1992], uncorrected) should approach 1 as runs converge. The
values may be unreliable if you have a small number of samples. PSRF should
only be used as a rough guide to convergence since all the assumptions
that allow one to interpret it as a scale reduction factor are not met in
the phylogenetic context.
Model parameter summaries over the runs sampled in files
"SmallData.bay.run1.p" and "SmallData.bay.run2.p":

The third thing to do is to find the table that summarizes the statistics for taxon bipartitions (**Figure 10.6**). Again, the values in the PSRF column should be near 1.0 if convergence has occurred.

Figure 10.6

Summary statistics for taxon bipartitions:

ID -- Partition	#obs	Probab.	Stddev(s)	Mean(v)	Var(v)	PSRF	Nruns
1 -- ....*....................	15004	1.000000	0.000000	0.067037	0.000219	1.000	2
2 -- ......*..........................	15004	1.000000	0.000000	0.239874	0.004529	1.003	2
3 -- ....*......................	15004	1.000000	0.000000	0.025169	0.000105	1.001	2
4 -- .....*...........................	15004	1.000000	0.000000	0.169111	0.001090	1.001	2
5 -- .....*...........................	15004	1.000000	0.000000	0.092149	0.000644	1.004	2
6 -- ....*******......	15004	1.000000	0.000000	1.001470	0.040379	1.000	2
7 -- ....*.....	15004	1.000000	0.000000	0.607192	0.014536	1.001	2
8 -- ....*........	15004	1.000000	0.000000	0.773123	0.016727	1.000	2
9 -- ....*..........	15004	1.000000	0.000000	1.120538	0.032732	1.000	2
10 -- ....*..........	15004	1.000000	0.000000	0.738751	0.019126	1.000	2
11 -- ....*.....	15004	1.000000	0.000000	0.764503	0.018886	1.000	2
12 -- ...*..........................	15004	1.000000	0.000000	0.174602	0.001517	1.002	2
13 -- ....*............	15004	1.000000	0.000000	0.412947	0.008944	1.002	2
14 -- ...***********************......	15004	1.000000	0.000000	0.221397	0.002023	1.002	2
15 -- .*********************************	15004	1.000000	0.000000	0.003661	0.000009	1.001	2
16 -- ..*...........................	15004	1.000000	0.000000	0.007936	0.000033	1.008	2
17 -- ....*....................	15004	1.000000	0.000000	0.072308	0.000321	1.001	2
18 -- ....*..................	15004	1.000000	0.000000	0.309398	0.002096	1.001	2
19 -- ....*..............	15004	1.000000	0.000000	0.160691	0.001028	1.000	2
20 -- ....*....................	15004	1.000000	0.000000	0.060074	0.000224	1.002	2
21 -- ....*......................	15004	1.000000	0.000000	0.173816	0.000705	1.001	2
22 -- ....*................	15004	1.000000	0.000000	0.070760	0.000258	1.001	2
23 -- ....*...............	15004	1.000000	0.000000	0.413526	0.005107	1.002	2
24 -- ..*...........................	15004	1.000000	0.000000	0.010864	0.000027	1.000	2
25 -- ....*......................	15004	1.000000	0.000000	0.241679	0.002228	1.071	2
26 -- ....*	15004	1.000000	0.000000	0.007120	0.000017	1.001	2
27 -- ....*..........	15004	1.000000	0.000000	1.037489	0.035360	1.003	2
28 -- ....*..............	15004	1.000000	0.000000	0.965386	0.029802	1.000	2
29 -- ....***...........	15004	1.000000	0.000000	0.520565	0.022162	1.015	2
30 -- ....*...........	15004	1.000000	0.000000	1.061894	0.033468	1.001	2
31 -- ....*......	15004	1.000000	0.000000	0.092818	0.002721	1.002	2
32 -- .....................*.*......	15004	1.000000	0.000000	0.002335	0.000005	1.004	2
33 -- ....**......	15004	1.000000	0.000000	0.776508	0.017602	1.000	2
34 -- ....*..........	15004	1.000000	0.000000	0.007987	0.000019	1.001	2
35 -- ....*........	15004	1.000000	0.000000	0.205139	0.002975	1.001	2
36 -- ....*...	15004	1.000000	0.000000	0.002152	0.000005	1.003	2
37 -- ......................*.*.	15004	1.000000	0.000000	0.002195	0.000005	1.000	2
38 -- ....*.............	15004	1.000000	0.000000	0.273842	0.002922	1.020	2
39 -- ........*...*.................	14989	0.999000	0.000094	0.062267	0.000301	1.006	2
40 -- .........*...*................	14866	0.990802	0.000754	0.043601	0.000284	1.000	2
41 -- ..........*...*..............	14771	0.984171	0.001070	0.060790	0.000620	1.002	2

In this case the chain swap frequencies are in the right range (see Figure 10.4), so the `temp` setting does not seem to be a problem. The PSRF values for the parameters are also in the right range (see Figure 10.5), and most of the bipartition PSRF values are in the right range (a few near the bottom of the table—not shown—are outside that range). The last value of the convergence diagnostic was 0.308 (see Figure 10.3), so convergence is questionable. The run required 1 hour and 25 minutes for 1,000,000 generations. My inclination would be to try again without changing any settings other than increasing `ngen` to 10,000,000 generations and let it run overnight.

As it happens, increasing `ngen` was all that was required for SmallData.bay to converge. Indeed, only 3,241,000 generations and 4 hours and 35 minutes were needed for convergence and all of the PSRF values in the two tables were in the 1.0-1.02 range.

But what if you never get convergence? Sometimes the data are just not sufficient to get convergence, which probably means they are not good enough to get a reliable tree. You may just have to live with that. The posterior probabilities of some nodes will probably reflect that unreliability (see *Learn More about Bayesian Inference*, p. 141).

What about Protein Sequences?

MrBayes can estimate trees from protein sequences quite nicely, but unless the coding sequences are unavailable there is little reason to do so. Estimating protein sequence trees takes several times longer than estimating the corresponding coding sequence trees. Templates for protein sequence analysis are included in the MrBayes Templates utility file.

Visualizing the MrBayes Tree

Although MrBayes draws crude trees in the `log.out` file, these are not suitable for publication. You will need to import the consensus tree into the tree-drawing program FigTree (see Chapter 1) to display the tree and possibly modify its appearance. By default MrBayes writes tree files in the FigTree format. If the input file was `SmallData.bay`, the file that contains the consensus tree will be `SmallData.bay.con`. The next section describes how to use FigTree to manipulate a tree drawing.

FigTree has most of the drawing features you are already familiar with from MEGA, but if you prefer to use MEGA to draw the tree simply open the `.con` file in FigTree, choose **Export Trees** from the **File** menu, and set the tree file format to **Newick**.

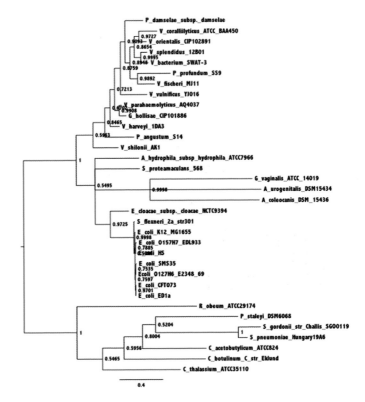

Figure 10.7

Figure 10.7 shows the Bayesian SmallData tree drawn by FigTree, with the posterior probabilities as node labels. The posterior probabilities are estimates of clade confidence and play the same role as bootstrap values do for NJ and ML trees. The posterior probabilities are calculated from the frequency with which a clade occurs among the trees that were summarized by the `sumt` command.

Notice that the clade consisting of *G. vaginalis*, *A. urogenitali* , and *A. coleocanis* is not part of a monophyletic Gram positive clade, as it was in the NJ and ML trees. Instead, 52% of the time that clade is found within the Gram negative bacteria. As a result, we cannot root the tree on the Gram positive organisms, so I instead rooted it on the Gram positive *R. obeum*–*C. thalassium clade*.

Figure 10.8 shows the LargeData BI tree as drawn by FigTree. Compare Figure 10.8 with the ML tree shown in Figure 9.13.

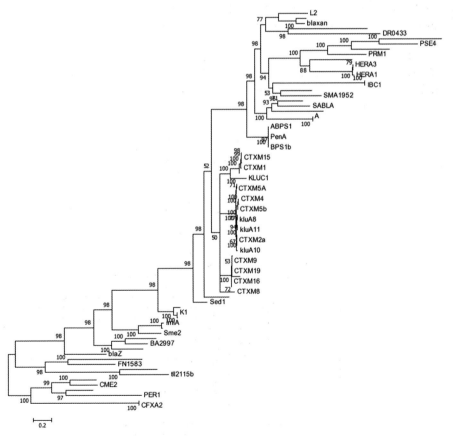

Figure 10.8

Using FigTree

FigTree can open tree files in its own extended version of Nexus format, in standard Nexus format, and in Newick format. When a file is opened, the tree layout is by default in the rectangular phylogram format (**Figure 10.9**). The layout and other aspects of the tree's appearance are modified by choices in the panel to the left of the drawing and by icons in the region above the drawing.

Figure 10.9

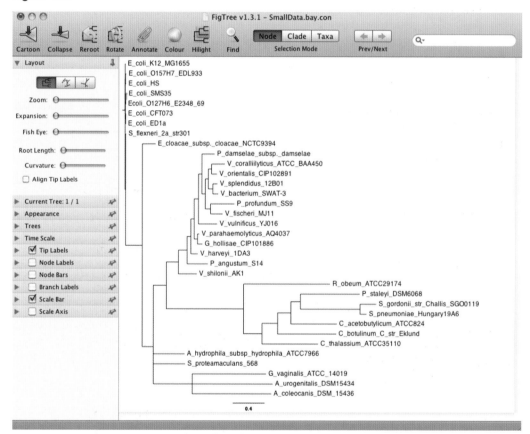

The side panel

There are 11 different aspects of the tree that can be modified in the side panel; each can be expanded by clicking the black triangle. By default the Layout section is expanded as in Figure 10.9. You should check out each of those sections to learn about the modifications it controls. Only a few are discussed here.

- **Layout:** The three icons allow you to choose among the *rectangular* format (default), the *circular* format, and the *radiation* (unrooted) format. The **Zoom** slider enlarges the view of the tree, while the **Expansion** slider enlarges it in the horizontal direction only. The **Curvature** slider makes the branches curve as they approach the node to the left. When moved to the extreme right, the curvature slider shows the tree in the slanted or straight format.

- **Trees:** The trees section (**Figure 10.10**) allows you root the tree at the midpoint by checking **Root Tree** and choosing **Midpoint** from the activated menu. Checking **Transform Branches** allows you to display the **Cladogram** format.

- **Branch Labels:** To show branch labels it is necessary both to tick the **Branch Labels** box and to choose what to display as a branch label (**Figure 10.11A**). I prefer **length_mean**. The **Sig. Digits** option (**Figure 10.11B**) lets you set the maximum number of digits that will be displayed, while the **Font** button allows you to set

Figure 10.10

Figure 10.11A

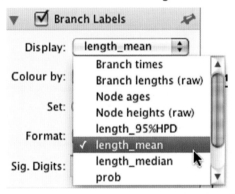

Figure 10.11B

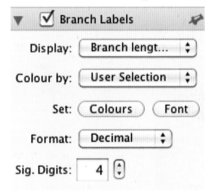

font and font size for the branch labels. If you set **Trees** to show the cladogram format, you will probably want to label the branches with lengths.

- **Node Labels:** Node labels works like branch labels. You must tick the box and choose what to display. For trees estimated by MrBayes, choose **prob** to display the posterior probabilities. (**Figure 10.12**).

Figure 10.12

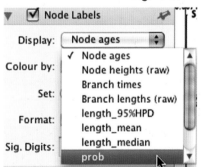

The icons above the tree

The icons above the tree in **Figure 10.13** are activated by clicking anywhere in the tree image.

- To *collapse* a clade to a triangle, click on the branch leading to the clade and click the **Collapse** or **Cartoon** icon. **Cartoon** leaves the taxon labels in place, while **Collapse** does not.

- To *reroot* the tree on an interior node, click on the branch leading to that node, then click the **Reroot** icon.

- To *rotate* a clade about a node, click on the branch leading to that node then click the **Rotate** icon.

Figure 10.13 shows the SmallData tree in cladogram format, rooted as in Figure 10.7, with both branches and nodes labeled. Notice that node labels (maximum 2 digits) are positioned to the right of the nodes, while branch labels (maximum 4 digits) are positioned above the branches.

Figure 10.13

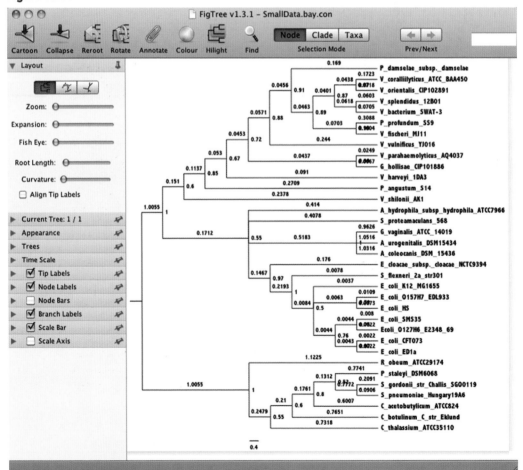

CHAPTER 11

Working with Various Computer Platforms

The first edition of *Phylogenetic Trees Made Easy* (Hall 2001) was justifiably criticized for being Mac-centric because it only discussed programs that ran on Macintosh operating systems. I tried to remedy that in the second edition (2004) by ensuring that every program I discussed was available for Macintosh, Windows, and Unix/Linux platforms. The third edition (2008) introduced MEGA 4. Because MEGA 4 was only available for the Windows operating system, the third edition was almost as Windows-centric as the first edition was Mac-centric. With the release of MEGA 5, which is available for Windows and Macintosh operating systems and will shortly be available for Linux, the fourth edition is truly platform-nonspecific. The other programs discussed in this book are likewise available for all operating systems.

Despite the best efforts of all concerned, however, programs still have platform-specific aspects. It is virtually impossible to modify every program so that it looks and behaves in exactly the way that is standard for each platform. If you are used to working in the Macintosh OS, a program that is imported from Windows may not look the way you expect it to and it may not be obvious how to do something. I have tried to address some of those issues in this chapter.

Command Line Programs

Most of the programs that are discussed use some form of the familiar Graphical User Interface (GUI), with menus, buttons, multiple function-specific windows, file-opening dialogs, and other operations all controlled by the mouse. GUI programs make it easy for the user but require a very sophisticated level of programming skills and tools. There are many valuable programs, all free, that use the much older command line interface in which every command is typed and the mouse does nothing at all. I discuss the basics of installing and using such programs later in this chapter.

MEGA on the Macintosh Platform

Written originally for Windows, MEGA's version 5 runs readily on Macintosh computers as well, but there are some adjustments you should be aware of. The list below offers a few quick caveats, followed by longer instructions involving folder management, navigation, and printing. Some of these instructions also apply to using MEGA in the Linux and Unix environments.

- *Starting MEGA.* MEGA takes around 30 seconds to start on a Mac because it is running within the Macintosh X11 program. If you check the Activity Monitor program while running MEGA you will notice that two programs, "mwine" and "wineserver," are running. The combination of X11, mwine and wineserver are the means by which MEGA 5, written for Windows, runs on the Mac.

- *Quitting MEGA.* Because MEGA 5 runs within the Macintosh X11 program, you cannot quit MEGA by typing the familiar `Command-Q`. The Command-Q keyboard shortcut is automatically disabled in X11 because it can interfere with some X11-based applications. You can terminate MEGA by choosing **Exit MEGA** from MEGA's menu, but X11 will still be running. To quit both MEGA and X11, choose **Quit X11** from the X11 menu at the top of the screen.

- *Alt and Control keys.* It would be impossible to rewrite all MEGA's Help files, tutorials, and manual to change the Windows terminology to Mac terminology. Mac users just need to be aware that when MEGA refers to the "Alt" key it means the Macintosh "Option" key. In most cases when MEGA mentions the Control key, either the Control or the Command key on the Mac will work, but the Control key always works.

- *Right-click.* Modern Macintosh mice have both right and left buttons, but the traditional Mac mouse had only one button, so "right-clicking" was accomplished by holding down the control key while clicking.

- *Opening MEGA files by double-clicking.* An odd feature of MEGA for Mac is that if MEGA is already running, double-clicking a MEGA file in the finder will not open the file; indeed nothing will happen. In contrast, if MEGA is *not* already running, double-clicking the file will start MEGA and open the file.

Navigating among folders on the Mac

The file-open dialog for MEGA always looks like that in Windows and may leave Mac users wondering how to find the folder they want. If confronted by a window that looks like **Figure 11.1**, double click the **My Documents** folder to open it.

Figure 11.1

The **My Documents** folder is your Home folder. The text box at the top of the window shows the name of the current folder (**Figure 11.2**). Clicking the button that looks like a folder with an arrow pointing up (arrow cursor in Figure 11.2) takes you up to the next level (in this case, back to Figure 11.1).

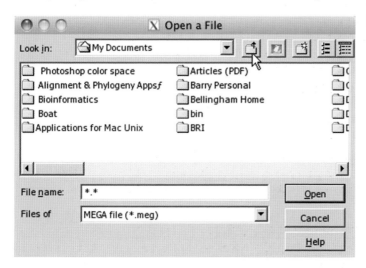

Figure 11.2

The button at the far right displays the contents of the current folder in a different format (**Figure 11.3**). The file dialog window cannot be enlarged, but the scroll bars allow you to see everything in the folder.

Figure 11.3

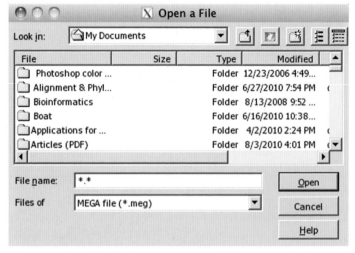

The **Files of** text box (**Figure 11.4**) determines what kind of files are shown, and the default is determined by what part of the program you are using. Choosing **All files** lets you see everything.

Figure 11.4

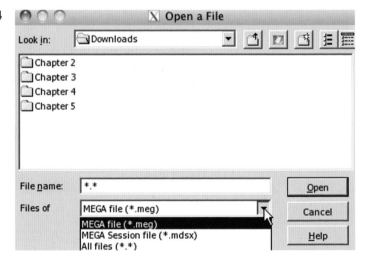

Printing trees and text from MEGA

MEGA for Mac does not support printing, but that is less of a problem that it would seem. The solution is to save whatever you want to print as a file, then print that file from another program. To print a tree save the file in PDF format from the **Image** menu of the **Tree Explorer** window, then open and print it from the Preview application that comes with the Mac OS. For most other windows, first save the window contents as a text file and then open and print it in your favorite text editor. An exception is the **Results** window when you choose **Find Best DNA/Protein Models (ML)** from the **Models** menu in MEGA's main window. That information is saved in HTML format, so you should give the file the extension `.html`, then open and print it from your favorite Web browser.

The Line Endings Issue

Files that are saved by MEGA all have Windows line endings, regardless of the platform MEGA is running on. Most of the time that doesn't matter because those files are used by MEGA itself. However, if you save a tree as a Newick file description, or save an alignment in FASTA or other format, with the intention of having a Macintosh or Linux program use that file you will probably need to change the line endings.

Windows, Macintosh, and Unix all use different codes to end lines. A file that is created in Windows has Windows (DOS) line endings and may not be read properly on a Macintosh or Linux machine. To make things more complicated, regular Macintosh applications need files with Macintosh termination, but files that are used by command line programs (i.e., those that run in the Terminal window) need Unix line termination.

Some programs are more picky than others. For instance, MEGA will draw trees from Newick tree description files that have Unix line termination without complaining. Just to be on the safe side, however, it is a good idea to ensure that all files have the proper termination for the operating system that will read them. This is not a problem for Windows users, who will probably do everything within the Windows environment, but Macintosh and Unix/Linux users need to be careful. Use your favorite text editor to convert a file to or from Windows or DOS line ending.

Installing Command Line Programs

MrBayes, codeml (see Chapter 14), and the various utility programs all normally require that the program reside in the same folder as its input files. If you have input files in many different folders this means continually moving the program from one folder to another. To avoid this serious inconvenience, put command line programs into a "bin" folder and provide a "path" to that folder. When that is done the command line programs can be run from within any folder that contains an input file, and any output files will be written to the same folder.

Macintosh and Linux: Use the bin folder

The best place to store command line programs is within the `/usr/local/bin` folder. I think it is best to store a *copy* of each command line program in the `/usr/local/bin` folder. That folder is normally invisible but you can reveal it by choosing **Go to folder...** from the Finder's **Go** menu and typing `/usr/local/bin` in the resulting dialog. Alternatively, you can drag the program file into `/usr/local/bin`. You will be told that the program can't be moved into the folder because `bin` can't be modified, but if you click the **Authenticate** button you will be asked for your name and password. Entering those allows a copy of the file to move. The "path" to that folder is already available to your system, so you do not need to create a path. **Note that you must have administrator privileges to install command line programs in the /usr/local/bin folder.** If you don't know what that means, get the administrator for your computer to install the programs for you.

Windows: Create a bin folder and a path to it

Windows does not have a "bin" folder already defined, so you must create one then create the path to it by the following steps:

1. Create a folder named `bin` directly on your C drive (i.e., not within any other folder).

2. Click **Start**, then right-click on **My Computer** and choose **Properties** from the menu to bring up the System Properties control panel (**Figure 11.5**).

Figure 11.5

3. Click the **Advanced** tab and at the bottom click the **Environmental Variables** button to see the dialog. In the **System variables** box at the bottom of the dialog, scroll down until you see **Path**. Select the **Path** item and click the **Edit** button (**Figure 11.6**).

Figure 11.6

4. Click in the **Variable value** box to deselect the path list and use the arrow key to move the cursor to the extreme right (**Figure 11.7**).

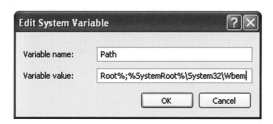

Figure 11.7

5. Type `;C:\bin` and *don't forget the semicolon* (**Figure 11.8**).

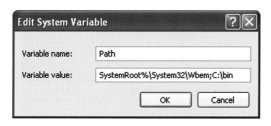

Figure 11.8

6. Click the **OK** button to close the dialog; click the **OK** button in the Environmental Variables window; and click the **OK** button in the System Variables window.

Put a copy of any command line program into that `bin` folder.

Command Line Programs: The Running Environment

Command line programs are run from within the **Terminal** program (Macintosh, Linux) or the **Command Prompt** program (Windows) by typing commands that are completed by hitting the **Return** or **Enter** key. What you type varies from program to program; the documentation that accompanies each command line program will describe the available commands and what they do.

Before discussing how to run the programs, a review of **Terminal** and **Command Prompt** is warranted.

Windows: A brief visit to the Command Prompt program

The Command Prompt program can be found in the **Accessories** folder under **All Programs** in the **Start** menu. It is convenient to create a shortcut to Command Prompt on the desktop.

When it is started the Command Prompt window looks like **Figure 11.9**. Its default size is 80 characters wide and it displays characters as white on a black background. You can right-click on the window title bar and choose **Properties** to modify the appearance of the window.

Figure 11.9

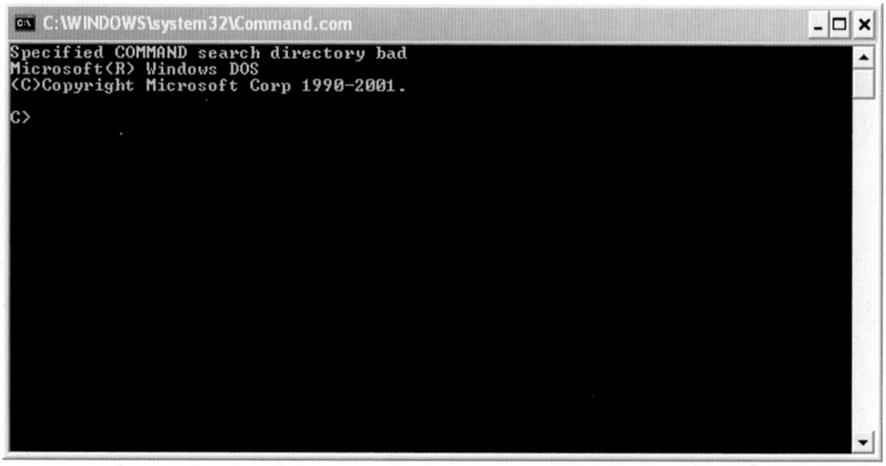

The **Layout** tab allows you to set the maximum width of the window (**Figure 11.10**). I like to set the **Screen Buffer Size** and **Window Size** widths to 120. The **Colors** tab allows you to set the background color of the window and the color of the characters to suit your taste.

Within the Command Prompt window the mouse does essentially nothing; all commands are entered by typing the command followed by the **Return** or **Enter** key.

In Figure 11.9, the bottom line (showing `C>`) is the **prompt**, which consists of the path to the *current directory* followed by `>`. Everything you type appears after the prompt. Programs within Command Prompt can only work on files in the current directory ("directory" is a synonym for "folder"), and all files are saved to the current directory. The current directory in Figure 11.9 is the C drive. The `cd` (change directory) command is used to make a different directory the current directory. Type `cd` followed by a space, then enter the *path* to the directory of interest. An easy way to enter that path is to drag the folder that you wish to make the current directory into the Command Prompt window. This will automatically enter the path to that folder. For instance, typing `cd` and then dragging a folder named "Perl Scripts" into the Terminal window changes the path to Perl Scripts (**Figure 11.11**). Making a particular directory the current directory is referred to as "navigating" to that directory.

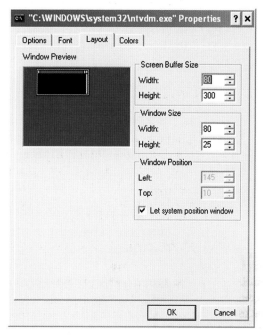

Figure 11.10

Figure 11.11

To see a list of the contents of a folder, type `dir`. The handy website http://www.computerhope.com/overview.htm provides a complete list of the DOS commands, but a discussion of most of those commands is beyond the scope of this book.

EXAMPLE USING THE COPY COMMAND   Suppose that you have a file named `MyProgram` in a folder named `UsefulStuff` and you want to copy that file to the `bin` folder that you created.

1. Start Command Prompt

2. Navigate to the UsefulStuff folder by entering `cd`, and then dragging the UsefulStuff folder into the Command Prompt window, then hitting the Return key.

3. Enter `copy MyProgram c:\bin` to copy the target file (`MyProgram`) to the destination folder (`c:\bin`)

4. Check to see that you successfully copied the MyProgram file to the bin folder

5. Navigate to the c:\bin folder by entering `cd c:\bin`

6. Enter `dir` to see a list of all the files in that folder. `MyProgram` should be in that list.

THINGS THAT MIGHT GO WRONG   You might get a message that says something like `The syntax of the command is incorrect`, which means that the target file you specified could not be found. One possibility is that the file name is actually `MyProgram v3.0` and you typed `copy MyProgram c:\bin`. **You must use the complete name of the target file.** Another possibility is that you typed `copy MyProgram v3.0 c:\bin`. The problem is that the `copy` command only recognizes the target file name up to the space, so it looked for a file named `MyProgram` and it did not find it. The solution is to surround the file name with double quotes; i.e. type `copy "MyProgram v3.0" c:\bin`.

Macintosh and Linux: A brief visit to Terminal and Unix

Terminal is a program that resembles a Unix terminal environment. The Terminal program can be found in the **Utilities** folder under **Applications**. It is convenient to put Terminal into the Dock (Mac). When you start Terminal, the screen will look something like **Figure 11.12**.

Figure 11.12

The line shown is the *prompt* and it ends with your short user name followed by the $ sign (or the % sign, depending on how your Terminal program is configured). Everything you type appears after the prompt, and you complete each command by hitting the **Enter** key. The part of the line before the short user name is the *path*, which is simply a description of the route to the folder that is the *current directory*. Unix uses the term "directory" instead of the term "folder," but they mean the same thing. When Terminal starts, the default setting for the current directory is your home directory; Unix uses the tilde (~) sign to indicate the home directory. In Figure 11.12, `Triton:~ barry$` means that Terminal is in the home directory of the user named `barry` on the drive named `Triton`. I will use `Courier` font to make it clear what you will type and what you will see in the directory window.

Unix programs can only operate on files that are in the current directory, and all files are saved to the current directory. To make a different folder the current directory type `cd` followed by a space, then enter the *path* to that directory. The easy way to enter the path is to type `cd` , then drag the folder of interest into the Terminal window. The complete path to that folder is displayed (**Figure 11.13**). Hit the **Enter** key to make that folder the current directory. Making a particular directory the current directory is often referred to as "navigating" to that directory. To see the contents of a directory, type `ls`.

Figure 11.13

EXAMPLE USING THE CP (COPY) COMMAND Suppose that you have a file named `MyProgram` in a folder named `UsefulStuff` and you want to copy that file to the `/usr/local/bin` folder.

1. Start the Terminal program.

2. Navigate to the UsefulStuff folder by entering `cd` , dragging the Useful-Stuff folder into the Terminal window, and then hitting the **Return** key.

3. Enter `sudo cp MyProgram /usr/local/bin` to copy the target file (My-Program) to the destination folder (/usr/local/bin).

4. You will be asked for your password. Enter that password then hit return.

5. Check to see that you successfully copied the MyProgram file to the /usr/local/bin folder

6. Navigate to the /usr/local/bin folder by entering `cd /usr/local/bin`

7. Enter `ls` to see a list of all the files in that folder. MyProgram should be in that list

What just happened in step 3? The directory `/usr/local/bin` is a special directory that is protected from casual access. In order to modify that directory by adding a file to it you must have administrator privileges on your account and you must be acting as the system's "root." It is very dangerous to act as the root because you can easily mess up your entire system. The `sudo` command allows you to be "root" for only the single command that follows the word `sudo`. That command was `cp`, which means to copy something. The `MyProgram` file is the "target" of the copy command and `/usr/local/bin` was the destination directory you were copying to.

THINGS THAT MIGHT GO WRONG   You might get a message like this one: `cp: No such file or directory`, which means that the file you listed could not be found. One possibility is that the file name is actually `MyProgram v3.0` and you typed `cp MyProgram /usr/local/bin`. **You must use the complete name of the file.** Another possibility is that you typed `cp MyProgram v3.0 /usr/local/bin`, which may be the correct file name. The problem is that the `cp` command only recognizes the name up to the space, so it looked for a file named `MyProgram` and it did not find it. The solution is to surround the file name with double quotes; i.e. type `cp "MyProgram v3.0" /usr/local/bin`.

Acquiring and Installing MrBayes

MrBayes 3.2 should be downloaded from http://www.mrbayes.net. It is available as a Windows program, a Macintosh application, or as source code that can be compiled for your own platform.

Windows users

MrBayes is a command line program that lacks the usual Windows interface, including navigation functions. `MrBayes.exe` must be run from within **Command Prompt**. To install MrBayes, move or copy `MrBayes.exe` to the `bin` folder you created.

Macintosh and Linux users

Macintosh users have two options. You can download the MrBayes application and run it like any other application, in which case MrBayes must be located in the same folder as the input file. If you use the MrBayes Macintosh application, the execution file must have Macintosh line endings and all output files will have Macintosh line endings. The Macintosh application will probably run somewhat more slowly than the binary that can be compiled from the source code as described below.

The better alternative is to download the source code for MrBayes, compile the source code into a binary, then install and run that binary as a command line program.

Compile MrBayes for your Mac

To compile the MrBayes source code, you need to have Developer Tools installed on your Mac. If you don't see a folder named **Developer** at the highest level of your hard drive, the Developer Tools are not installed. The install disc that came with your Mac will have a folder named **Xcode Tools**. Open that folder and double-click the package that says **Xcode Tools.mpkg** to install Developer Tools. Do a restart and you are ready to go. You need not ever open the Developer Tools folder to work with it.

Download the source code folder as described on the MrBayes website. Follow the instructions for compiling MrBayes. Briefly, they are:

1. In Terminal, navigate to the source code folder.

2. *Create a configure file.* Enter `autoconf`. A configure file will be created within the `source code` folder.

3. *Create the makefile file.* Enter `./configure`. A file named Makefile will be created within the `source code` folder. Check to be sure that the file has content (size is > 0 kb and there is text if you open it in your favorite text editor).

4. *Compile MrBayes.* Enter `make`. The Unix executable file `mb` will be created.

5. *Get rid of the object files.* Enter `rm *.o` (that's an "oh", not a zero).

6. *Install MrBayes.* Enter `sudo cp mb /usr/local/bin`. **You will need your administrator password for this step.**

7. Close the Terminal window.

You will need to open a new Terminal window to use MrBayes.

Running the Utility Programs

The Utility programs are two programs that I have written in the Perl language to make some tasks easier (**Table 11.1**). The programs have been compiled as executables for Windows, Macintosh, and Linux. The programs are:

- A file format conversion program, **FastaConvert**. Converts alignments in Fasta format to Nexus and other formats.

- An ancestral sequence estimation program, **ExtAncSeqMEGA**. Extracts ancestral sequences from a file written by MEGA (see Chapter 14).

It is your responsibility to ensure that the input files for the Utility programs have the correct line endings for your platform. In particular, files that have been created by MEGA, *whatever the platform MEGA is run on*, will have Windows line endings. If you are using a Macintosh or Linux operating system be sure to change the line endings to Unix style.

TABLE 11.1 Downloadable Utility Programs[a]

FASTACONVERT	
Function	Converts an alignment in FASTA format to one of several other formats
Input file	An alignment in FASTA format. Each sequence can be on a single line or on multiple lines
Usage command	`FastaConvert myfile.fas [-n -f -p -x]`[b]
Output	An alignment in the specified format with the extension .nxs, .fas, .phy, or .pam[b]
EXTANCSEQMEGA	
Function	Extracts ancestral DNA and protein sequences from the **Detailed Text Export** file written by MEGA (see Chapter 14)
Input file	The **Detailed Text Export** file written by MEGA
Usage command	`ExtAncSeqMEGA myfile.anc [cutoff]`[c]
Output	A set of files, one for each internal node, that give the ancestral sequences, their accuracy scores and their probabilities. The files are collected into a folder named **myfile.anc output files**.

[a] Programs can be downloaded from http://sites.sinauer.com/hall4e/. See text for instructions.

[b] The first argument is the name of the input FASTA alignment file. The arguments enclosed in square brackets specify the format(s) of the output alignment file. At least one format must be specified, e.g., –n. Multiple output formats can be specified in any order, but they must be separated by spaces. Do not type the square brackets.
-n = Nexus format = .nxs
-f = FASTA format = .fas (in which each sequence occupies exactly one line)
-p = PHYLIP format modified to permit taxon names up to 50 characters = .phy
-x = PHYLIP format required by the PAML suite of programs = .pam

[c] [cutoff] is an optional argument that sets the ratio of MPAA to MPAA2, below which an amino acid is noted as being unreliable. The default value is 1.5. Higher values are more stringent. Do not type the square brackets.

Utility programs for Windows

The utility programs are in the Chapter 11 folder in the Windows package, which you can download from the website http://sites.sinauer.com/hall4e/. Copy or move these from the utility programs folder to the `bin` folder that you created.

The utility programs are run from within the Command Prompt program. Make the folder that contains the input file for that particular program the current directory as described above. Type the name of the program plus any required parameters. For example, to run the program ExtAncSeqMEGA with the input file name `myfile.anc` you must type `ExtAncSeqMEGA myfile.anc`. **Remember! The input file must have Windows line endings.** Details about running the programs, including any required or optional parameters, are provided in Table 11.1.

Utility programs for Macintosh and Linux

The utility programs are in the Chapter 11 folder in the Macintosh package, which you can download from the website http://sites.sinauer.com/4hall4e/.

You must have Administrator privileges to install a program. In the Terminal window, navigate to the Chapter 11 folder containing the utility programs. Type `sudo cp ApName /usr/local/bin/` (where the program is named `Ap-Name`). You will be asked for your Administrator password.

That's it. You have copied the executable into the invisible folder called `usr/local/bin`. You can check to see that the copy was successful by typing two commands, `cd /usr/local/bin/` and `ls`. You will see a list of the items in `bin`, and the program should be among them. Close the Terminal window to complete the installation.

You are now ready to use the program, but first you must open a new Terminal window by typing `Command-N`. A window checks `/usr/local/bin/` for all the commands it can execute when it first opens, so the window you used to install the program won't recognize that the program is installed.

In Terminal, make the folder that contains the input file for a program the current directory. **Remember! All input files must have Unix line termination.** Type the name of the program followed by any required parameters. For example, to run the program ExtAncSeqMEGA with the input file named `myfile.anc` type `ExtAncSeqMEGA myfile.anc`.

Warning: Terminal runs in a Unix environment, and Unix hates spaces—it interprets spaces as the ends of commands or arguments. For that reason, naming an input file `my file.fas` is a very bad idea. If you type the command `FastaConvert my file.fas -n`, FastaConvert will look for an input file named `my`, won't find it, and will complain and quit. To make file names readable, use `myFile.fas` or `my_file.fas`.

CHAPTER 12

Advanced Alignment Using GUIDANCE

Chapter 4 discussed the importance of alignment accuracy to phylogenetic tree estimation, but also pointed out that tree accuracy is surprisingly tolerant of alignment inaccuracy (Ogden and Rosenberg 2006). Other applications of MSAs, however, are quite sensitive to alignment accuracy. Those applications include reconstruction of ancestral sequences (Chapter 13) and detection of adaptive evolution (Chapter 14). For those applications, it is essential to maximize the accuracy of the alignment. GUIDANCE is a Web-based program that creates multiple sequence alignments (MSAs), then evaluates the reliability of those alignments.

Just how does one determine alignment accuracy? Various studies have compared accuracies of alignment methods by using databases of "gold standard," structure-based alignments of proteins and by using simulations, but that doesn't help much when it comes to your specific alignment of your data. Alignment accuracy is very much affected by the data. Inclusion of a nonhomologous sequence can distort the alignment of the other sequences, resulting in an overall low-quality alignment. Some regions of the sequences may have diverged so much that they align badly, likewise distorting the remainder of the alignment. There simply is no intuitive way to detect such situations.

Chapter 4 suggests using the average amino acid identity as a guide to the reliability of alignments, but does not suggest what to do about alignment of non-coding sequences or alignment of RNAs.

We really need to know two things. First, is my alignment reliable? And second, if it isn't, what can I do about it? GUIDANCE offers answers to both questions.

Issues of Alignment Reliability

Unreliable sequences

Unreliable sequences are often included as the result of seeking outgroup sequences with which to root a tree. We know that we need one or more outgroup sequences to root a tree (see Chapter 7), and we know that an outgroup sequence is more distantly related to the ingroup sequences than the ingroup sequences are to each other. We also know that we need some "outside information" to choose an outgroup sequence, information that doesn't come from the sequences themselves. For instance, we might use an insect sequence to root a tree in which the ingroup sequences came from mammals. The problem is one of finding a sequence that is sufficiently distant, but not so distant as to distort the alignment. GUIDANCE detects the sequences that contribute to unreliability and offers an automatic way to redo the alignment without those sequences. In this chapter we will look at an example case in which the inclusion of some sequences that align unreliably distorts a tree.

Unreliable regions

The portions of sequences that encode amino acids on the interior of proteins tend to be more conserved than do the regions that encode loops on the surface of proteins. As a consequence, surface region residues tend to align less reliably than do those of the interior regions. We will also consider the effects of unreliable regions on detecting adaptive evolution.

How GUIDANCE Works

Multiple alignment programs, one way or another, use a "guide tree" to iteratively add sequences to a multiple alignment. Those guide trees contribute a significant source of error to the final alignment (Penn et al. 2010). GUIDANCE uses a bootstrap approach to estimate the reliability of the individual columns in that final alignment, a procedure that is comparable to assessing the reliability of different nodes in a tree by bootstrap sampling. GUIDANCE first estimates a "base" alignment either by the MAFFT, the ClustalW, or the PRANK program. It then estimates a bootstrap NJ tree from that base alignment. Recall that each bootstrap sample involves sampling random sites from the base alignment to make a "pseudoalignment," then estimating a "pseudotree" from that pseudoalignment. GUIDANCE uses each of those pseudotrees as a guide tree to make an MSA from the unaligned sequences. Those MSAs differ from each other and from the base alignment only in being based upon different guide trees.

Finally, GUIDANCE calculates, for each residue pair in the base alignment, the fraction of pseudoalignments in which that residue pair was aligned together. The result is a GUIDANCE score for each residue pair in the base alignment. A high score (i.e., near 1.0) means that the pair aligns very reliably regardless of the initiating guide tree, while a low score means that the position of the residue in the alignment is sensitive to the guide tree. The score for each column in the alignment is the average of the scores for the residue pairs in that column. Those scores reflect the confidence in the alignment at each site. The method is described in more detail in Penn et al. 2010.

Naturally, doing 100 or more bootstrap alignments takes considerable computer time, and aligning 100 sequences typically takes about an hour when using MAAFT.

An Example Illustrated by the SmallData Data Set

The NJ, Parsimony, ML, and BI trees we generated from the SmallData data set were rooted by using the Gram positive organisms as an outgroup. In the NJ, Parsimony, and ML trees (see Figures 7.18, 8.9, and 9.12), the Gram positive organisms formed a monophyletic group that could be used to root those trees. The BI tree, however, put three of those Gram positive organisms within one of the Gram negative clades (**Figure 12.1**; see also Figure 10.7). For this example we will look at the reliability of the sequences in that alignment.

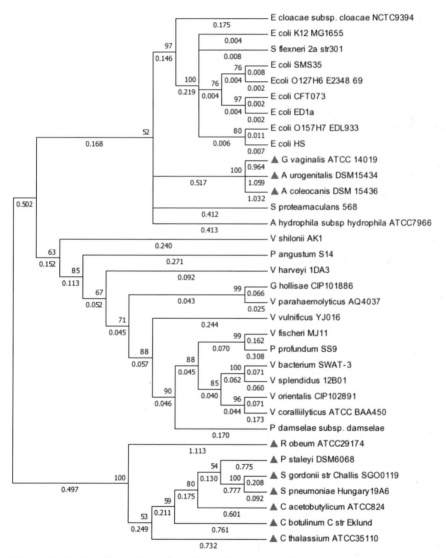

Figure 12.1 SmallData BI tree based on all sequences. Gram positive taxa are indicated by a red triangle.

Make a file of the unaligned sequences in FASTA format

GUIDANCE requires as its input a set of sequences in the FASTA format. In MEGA's main window choose **Edit/Build Alignment** from the **Align** menu, then choose **Open a saved alignment session** to open a saved alignment in the `.mas` format. To unalign the sequences, first select all of the sequences by typing `control-A`, then choose **Delete Gaps** from the **Edit** menu. Finally, choose **Export Alignment** from the **Data** menu and choose **FASTA format** in the submenu. I typically add the letter "U" to the name of the alignment to remind myself that these are unaligned sequences.

 Chapter 12: SmallDataU.fas

Starting the run

In your favorite web browser go to http://guidance.tau.ac.il/. Click the **Browse** button (indicated by cursor arrow in **Figure 12.2**), navigate to the folder that contains the saved FASTA file, and open that file. It is better to do this than to paste the sequences into the text box because that way GUIDANCE reports the name of the input file in its results output—a handy reminder for the future.

From the **Sequences Type** choices, choose **Codons** if the sequences are DNA coding sequences, **Nucleotides** for other DNA or RNA sequences, or **Amino acids** for protein sequences.

Use the drop-down menu to **Select the MSA algorithm**. The choices are **MAFFT (default)**, **PRANK** and **ClustalW**. MAFFT is the fastest method and seems to be significantly more accurate than ClustalW (Nuin et al. 2006). PRANK is a newer method that is reported to be much more accurate than either MAFFT or ClustalW in terms of gap placement (Loytynoja and Goldman 2008), but it is much slower than either MAFFT or ClustalW. Note that for the Codons option the sequence must not include any non-coding region, and thus must be of a length dividable by three.

Be sure to enter your email address in the text box. GUIDANCE will send you an email notification when your alignment is ready, and that notification includes a link that will let you get back to your alignment on the GUIDANCE server for a period of three months.

Finally, enter a job title—something descriptive that will remind you exactly what this job was.

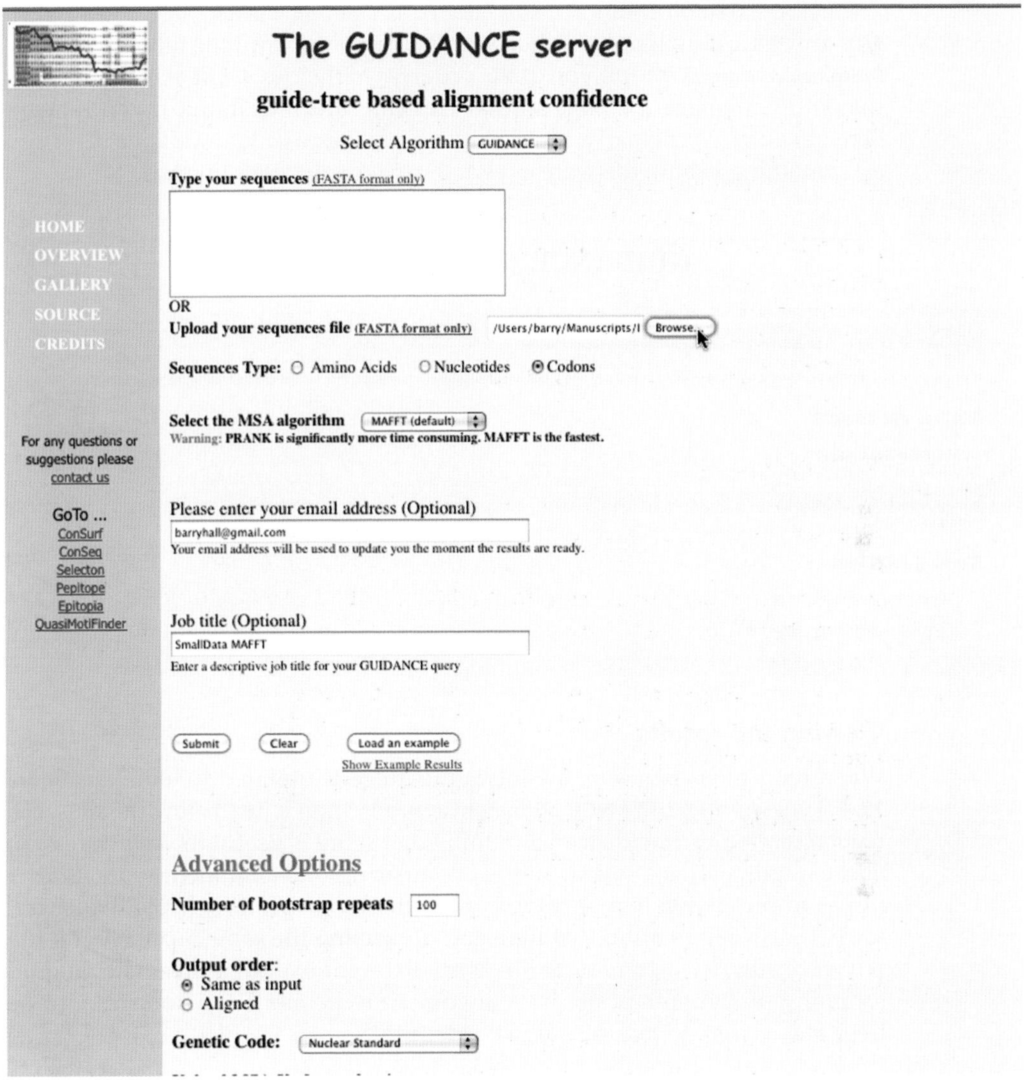

The GUIDANCE server

guide-tree based alignment confidence

Select Algorithm [GUIDANCE ‡]

Type your sequences (FASTA format only)

OR

Upload your sequences file (FASTA format only) /Users/barry/Manuscripts/I [Browse...]

Sequences Type: ○ Amino Acids ○ Nucleotides ⊙ Codons

Select the MSA algorithm [MAFFT (default) ‡]
Warning: PRANK is significantly more time consuming. MAFFT is the fastest.

Please enter your email address (Optional)
barryhall@gmail.com
Your email address will be used to update you the moment the results are ready.

Job title (Optional)
SmallData MAFFT
Enter a descriptive job title for your GUIDANCE query

[Submit]   [Clear]   [Load an example]
Show Example Results

Advanced Options

Number of bootstrap repeats [100]

Output order:
⊙ Same as input
○ Aligned

Genetic Code: [Nuclear Standard ‡]

HOME
OVERVIEW
GALLERY
SOURCE
CREDITS

For any questions or
suggestions please
contact us

GoTo ...
ConSurf
ConSeq
Selecton
Pepitope
Epitopia
QuasiMotiFinder

Figure 12.2

If you have chosen the ClustalW algorithm (which is *not* recommended), you might want to scroll down to the bottom of the page to the **Advanced Options** section and change the **Output Order** from the default **Aligned** to **Same as input**. Otherwise the output alignment is written in the order in which the sequences are added to the alignment, which can make it difficult to locate a particular sequence if there are a lot of sequences.

Click the **Submit** button to start the run. The **Job Status** page (**Figure 12.3**) will appear and will be refreshed every 30 seconds until the job is done. The **Running Messages** inform you of the progress of the run. GUIDANCE suggests that you bookmark the page so that you can return to it, but if you remembered to enter your email address that is not necessary. Just walk away and go do something useful; the watched GUIDANCE job never finishes.

Figure 12.3

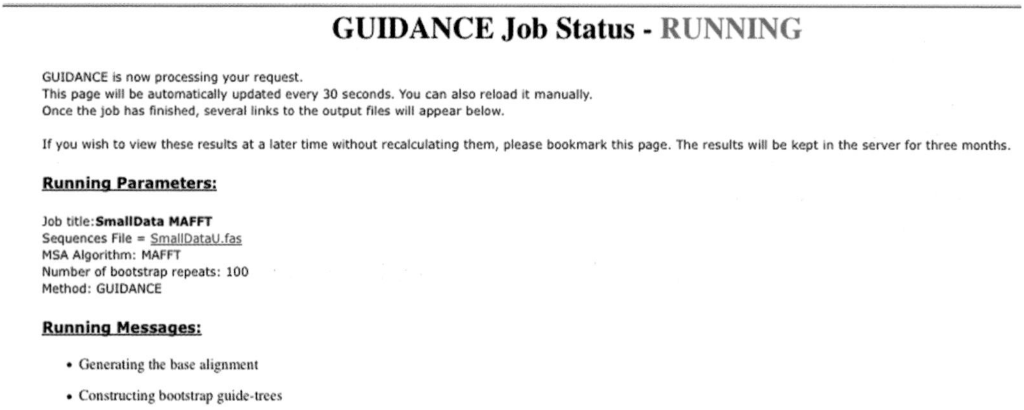

GUIDANCE Job Status - RUNNING

GUIDANCE is now processing your request.
This page will be automatically updated every 30 seconds. You can also reload it manually.
Once the job has finished, several links to the output files will appear below.

If you wish to view these results at a later time without recalculating them, please bookmark this page. The results will be kept in the server for three months.

Running Parameters:

Job title:**SmallData MAFFT**
Sequences File = SmallDataU.fas
MSA Algorithm: MAFFT
Number of bootstrap repeats: 100
Method: GUIDANCE

Running Messages:

- Generating the base alignment

- Constructing bootstrap guide-trees

Viewing the results

When the run is complete, the Job Status page is updated to look like **Figure 12.4**. The overall **GUIDANCE alignment score** is shown just below the **MSA colored according to confidence score** link. That score is useful for an impression of the alignment as a whole and for comparing one alignment with another (for example, an alignment prior to and after removing unreliable sequences; see p. 186). **Keep in mind that repeated runs using the same input FASTA file will give slightly different results with different overall scores.** The random sampling of columns in the base alignment to generate the bootstrap guide trees means that no two runs will be identical.

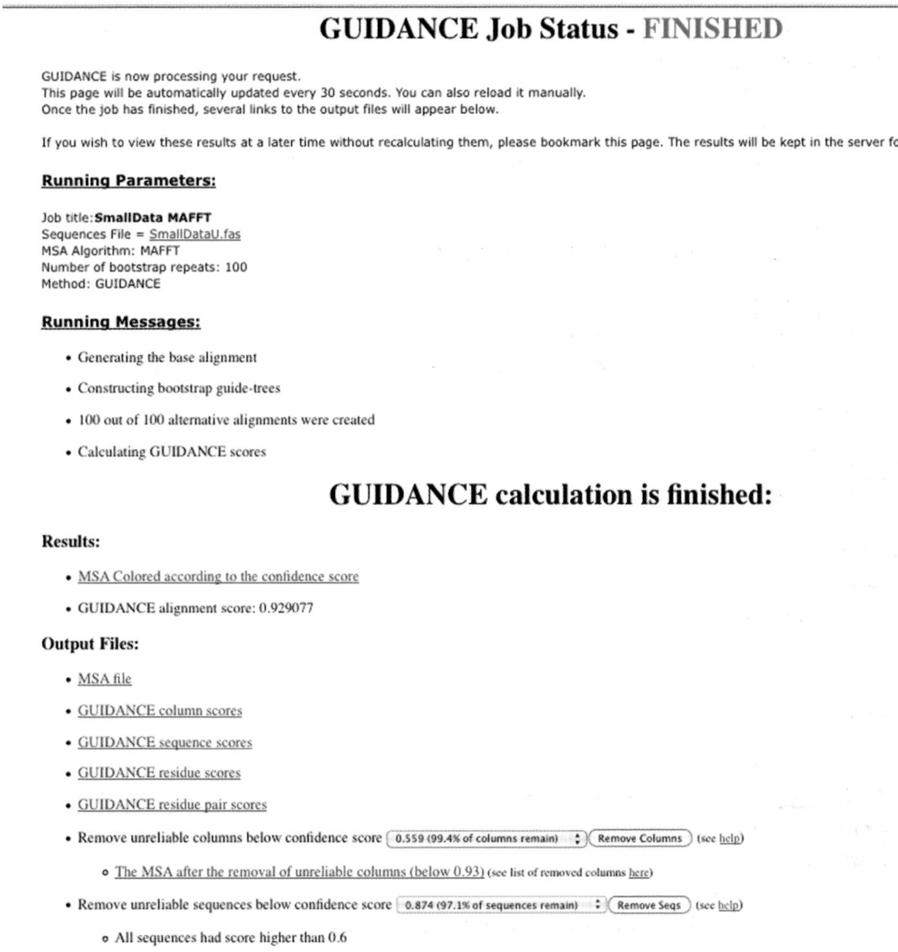

GUIDANCE Job Status - FINISHED

GUIDANCE is now processing your request.
This page will be automatically updated every 30 seconds. You can also reload it manually.
Once the job has finished, several links to the output files will appear below.

If you wish to view these results at a later time without recalculating them, please bookmark this page. The results will be kept in the server for three months.

Running Parameters:

Job title:**SmallData MAFFT**
Sequences File = SmallDataU.fas
MSA Algorithm: MAFFT
Number of bootstrap repeats: 100
Method: GUIDANCE

Running Messages:

- Generating the base alignment

- Constructing bootstrap guide-trees

- 100 out of 100 alternative alignments were created

- Calculating GUIDANCE scores

GUIDANCE calculation is finished:

Results:

- MSA Colored according to the confidence score

- GUIDANCE alignment score: 0.929077

Output Files:

- MSA file

- GUIDANCE column scores

- GUIDANCE sequence scores

- GUIDANCE residue scores

- GUIDANCE residue pair scores

- Remove unreliable columns below confidence score [0.559 (99.4% of columns remain) ⬍] (Remove Columns) (see help)

     o The MSA after the removal of unreliable columns (below 0.93) (see list of removed columns here)

- Remove unreliable sequences below confidence score [0.874 (97.1% of sequences remain) ⬍] (Remove Seqs) (see help)

     o All sequences had score higher than 0.6

Figure 12.4

Figure 12.5

Legend:

The alignment confidence scale:

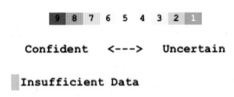

Confident   <--->   Uncertain

Insufficient Data

Clicking the **MSA colored according to confidence score** link under **Results** displays the base alignment in a new window, which as the name says is color-keyed by confidence level. **Figure 12.5** displays the 5′ end of that alignment and shows that most of the visible residues have very high scores (see the Legend, left). Below the alignment is a series of blue bars whose height indicates the average score for that column in the alignment. Putting the cursor over such a bar displays the column number followed by the average score for that column. The colored

MSA color-coded by GUIDANCE scores

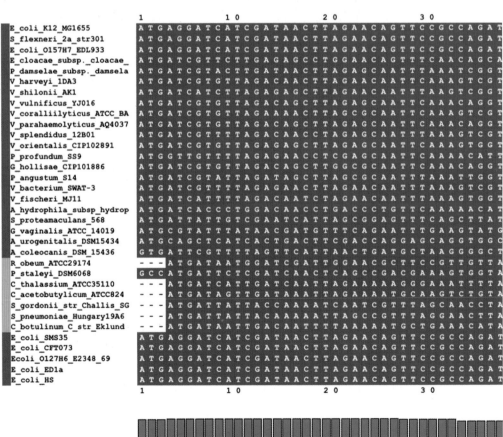

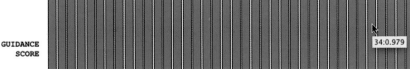

column to the left of the sequence names shows the average score for the residues in that sequence using the same colored confidence scale.

Using the scroll bar at the bottom of the window, we can scroll over to see the rest of the alignment. **Figure 12.6** shows a region of generally low confidence.

Figure 12.6

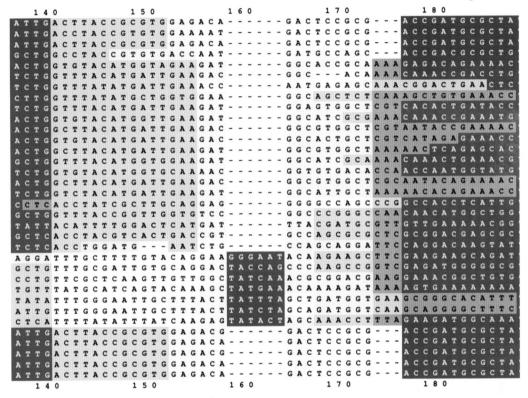

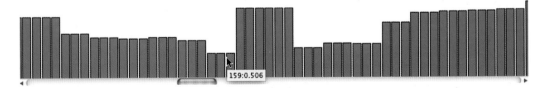

Below the results section on the Job Status page (see Figure 12.4) are links to five **Output Files**, including:

- **MSA File**, the base alignment in FASTA format.

- **GUIDANCE column scores**, which is a list of the average score for each column—i.e., the number that is displayed in the bar graph below the alignment in the colored display.

- **GUIDANCE sequence scores,** a list of the average score for each sequence. You need to look at this list to see the exact score for each sequence.

Each of the output files is viewed by clicking the link. Each file can be saved as a text-only file. If the MSA file is saved with the extension `.fas` it can be imported back into MEGA or it can be converted to another format (such as Nexus) with FastaConvert (see Chapter 11).

Eliminate unreliable sequences

Figure 12.5 shows that seven of the sequences have lower reliability than the rest (pink). All seven are from Gram positive organisms. In contrast, the Gram positive sequences that are misplaced in the BI tree (see Figure 12.1) have high reliability. We will see what effect removing the lower reliability sequences has on estimating the BI tree.

Near the bottom of the Job Status window are two buttons that allow you to remove unreliable columns or sequences (**Figure 12.7**).

Figure 12.7   • Remove unreliable columns below confidence score [0.559 (99.4% of columns remain) ⬍] (Remove Columns) (see help)

 ○ The MSA after the removal of unreliable columns (below 0.93) (see list of removed columns here)

 • Remove unreliable sequences below confidence score [0.874 (97.1% of sequences remain) ⬍] (Remove Seqs) (see help)

 ○ All sequences had score higher than 0.6

Figure 12.8

```
SEQUENCE_NAME     SEQUENCE_SCORE
E_coli_K12_MG1655          0.939
S_flexneri_2a_str301       0.939
E_coli_O157H7_EDL933       0.939
E_cloacae_subsp._cloacae_NCTC9394      0.938
P_damselae_subsp._damselae        0.948
V_harveyi_IDA3  0.947
V_shilonii_AK1  0.935
V_vulnificus_YJ016         0.943
V_coralliilyticus_ATCC_BAA450     0.948
V_parahaemolyticus_AQ4037         0.949
V_splendidus_12B01         0.947
V_orientalis_CIP102891  0.948
P_profundum_SS9 0.947
G_hollisae_CIP101886       0.949
P_angustum_S14  0.946
V_bacterium_SWAT-3         0.947
V_fischeri_MJ11 0.947
A_hydrophila_subsp_hydrophila_ATCC7966  0.917
S_proteamaculans_568       0.945
G_vaginalis_ATCC_14019     0.935
A_urogenitalis_DSM15434  0.902
A_coleocanis_DSM_15436     0.925
R_obeum_ATCC29174          0.873
P_staleyi_DSM6068          0.885
C_thalassium_ATCC35110     0.858
C_acetobutylicum_ATCC824          0.883
S_gordonii_str_Challis_SG00119    0.882
S_pneumoniae_Hungary19A6          0.882
C_botulinum_C_str_Eklund          0.885
E_coli_SMS35    0.939
E_coli_CFT073   0.939
Ecoli_O127H6_E2348_69      0.939
E_coli_ED1a     0.939
E_coli_HS       0.939
#END
```

To the left of each button is a drop-down menu that lets you choose the cutoff for removing columns or sequences. Figure 12.4 showed that the sequences all have scores in the range of 0.8 to 1.0. To see the exact scores for each sequence (**Figure 12.8**), click the **GUIDANCE sequence scores** link in Figure 12.4. All of the suspect sequences have scores less than 0.902.

From the dropdown menu to the immediate left of the **Remove Seqs** button (**Figure 12.9**), choose the first score that is higher than that of the sequence to be removed (in this case 0.902), then click the **Remove Seqs** button.

Figure 12.9

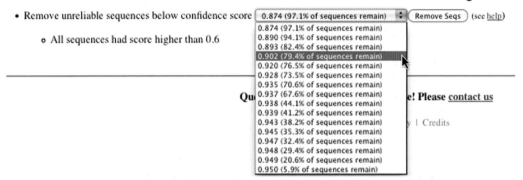

A new link appears along with a button that allows you to run GUIDANCE on the confidently aligned sequences only (**Figure 12.10**). Clicking the **here** link (red arrow) shows a list of the sequences that were removed. Clicking the **run GUIDANCE on the confidently-aligned sequences only** button brings up a new data entry page for GUIDANCE with the sequences already in place and the same setting used in the original run. I always enter a new job title to distinguish this from the original run, in this case `SmallData MAFFT reliable sequences`, and re-enter an email address for notification. We can then use the result of that run to proceed to the next stage. In this case, however, we will simply go to the next stage without eliminating the sequences.

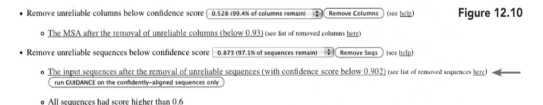

Figure 12.10

The result of aligning just the reliable sequence is an overall alignment score of 0.994, as opposed to 0.929 when all of the sequences were aligned. **Figure 12.11** is the same region of the alignment as was shown in Figure 12.6. Note that, with minor exceptions, all residues are reliably aligned. Also note that the column that had a score of 0.506 in Figure 12.6 has a score of 0.985 in Figure 12.11. (The columns are numbered differently because there are fewer gaps after removing the unreliable sequences.) The presence of the unreliable sequences clearly perturbed the reliability of this region, even for the reliable sequences.

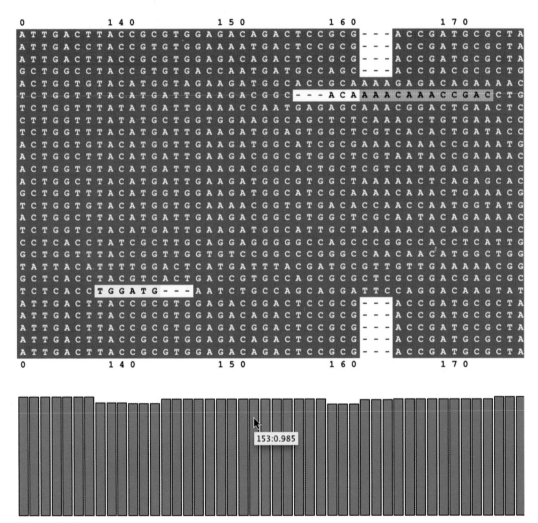

Figure 12.11

To determine the effect of removing the unreliable sequences on the BI tree, I estimated the BI tree from the MAFFT alignment of the complete SmallData set (**Figure 12.12A**) and from the MAFFT alignment of the reliable sequences (**Figure 12.12B**).

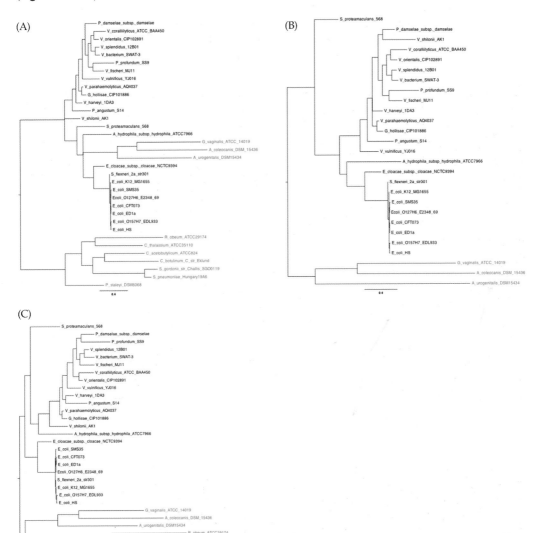

Figure 12.12 Gram positive taxa are shown in red. (A) SmallData BI tree based on all sequences. (B) SmallData BI tree based on reliably aligned sequences only. (C) SmallData BI tree based on reliably aligned columns only of all sequences.

The BI tree based on only the reliable sequences shows that the Gram positive organisms form a monophyletic group that can be used to root the tree (see Figure 12.12B). On the other hand, the rest of the Gram positive sequences were also eliminated as being unreliable, so it is not clear that we really gained much beyond having a more reliable alignment. An alternative would be to keep all of the sequences and eliminate the unreliable *regions* of the alignment instead.

How can we decide an appropriate cutoff score for eliminating columns? Notice that GUIDANCE has already calculated an MSA after removal of unreliable columns below the default score of 0.93 (see Figure 12.7). That default value was chosen to reflect the best balance between false negative errors (i.e., correctly aligned columns that are marked as unreliable) and false positive errors (i.e., badly aligned columns that are marked as reliable). However, this cutoff will have very different effects on different data sets that may display variable levels of alignment uncertainty. We can use the drop-down menu to determine that 63.9% of the columns will remain in the case of the SmallData example. That doesn't seem terribly unreasonable. There is no simple rule for how much data are enough to estimate a good phylogeny, partly because it depends on whether the remaining data include sufficient variation. My own rule of thumb is that I feel safe if at least 50% of the data remain. In any event, the ultimate test will be the confidence in the eventual tree (posterior probabilities or bootstrap values).

To accept the default MSA after removal of the unreliable columns, click the link for the trimmed MSA to display it in FASTA format. From your browser's **File** menu, save the file in **Text** format with the extension `.fas`.

Figure 12.12C shows the BI tree based on the reliable columns from all of the sequences. With the unreliable columns removed, the Gram positive sequences form a monophyletic group that can be used to root the tree. Compare Figure 12.12C with Figures 7.18, 8.9, and 9.12. For this data set at least, BI appears to be more sensitive to unreliably aligned bases than are NJ, Parsimony, or ML.

Applications of GUIDANCE

This chapter has illustrated using GUIDANCE to improve the estimation of phylogenetic trees, probably the least important use of GUIDANCE. Two applications for which GUIDANCE is more important—estimation of ancestral sequences and detection of adaptive evolution—are discussed in Chapters 13 and 14, respectively. The use of GUIDANCE may also be important in other kinds of analyses that rely on an alignment, such as the detection of horizontal gene transfer or site-specific coevolution.

Removing columns is not an option if you intend to use the data set to estimate ancestral sequences as described in the next chapter. In that event you must retain all columns, but you should keep in mind that columns that are aligned unreliably will probably produce unreliable estimates of ancestral sequence in those regions.

Reconstructing Ancestral Sequences

Ancestral sequence reconstruction is similar to phylogenetic tree reconstruction: we cannot *know* the sequences of ancestral proteins, but we can *estimate* those sequences. Once we have an estimate of the sequence of an ancestral protein, we can synthesize a gene that encodes that sequence. We can then express that gene to obtain the physical protein. With the protein in hand we can determine its biochemical properties and activities. From these properties we may be able to draw inferences about cellular and environmental conditions that existed in the past.

Ancestral sequence reconstruction always begins with estimation of a phylogenetic tree. We then estimate the sequence of a gene or protein, which is represented by an internal (hence ancestral) node on that tree. For instance, if we know that β-glactosidases and β-glucuronidases are descended from a common ancestor, we might like to know whether that common ancestor acted on β-galactoside substrates, on β-glucuronide substrates, or on both (or perhaps on neither). After these proteins diverged, what were the specificities of the enzymes at the first nodes following divergence? Were those ancient enzymes much more general glycosidases than the enzymes existing today? It is possible to answer such questions with a reasonable degree of confidence.

Most of this discussion focuses on ancestral protein sequence reconstruction, but ancestral DNA and encoded RNA sequences may also be of interest. If you are estimating ancestral RNA sequences you will need to pay special attention to the alignment stage, taking paired stems and loops into consideration. The methods discussed here all involve first estimating the ancestral DNA sequence, then estimating the ancestral protein sequence; those interested in ancestral RNA sequences can simply ignore everything that is said about proteins *per se*.

The first step in ancestral sequence reconstruction is to make an alignment of the genes that encode the set of existing proteins of interest. **It is essential that the DNA sequence alignment be based on alignment of the corresponding protein sequences** as described in Chapters 4 and 12. The more distantly related the sequences are, the more gaps will be present in the alignment, with the result that the overall alignment typically is 30% to 70% longer than any of the sequences it contains. Until recently, ancestral reconstruction methods ensured that the estimated ancestral sequence were exactly as long as the alignment and were without gaps. Thus, except for highly conserved or very recently diverged sequences, the estimated sequences were unrealistic, making it impractical to physically reconstruct the proteins and obtain any useful information.

This aspect of ancestral state reconstruction has greatly limited the general applicability of what some have called "molecular paleontology." A method to overcome this problem has been described, however. By correcting the estimated sequences for ancestral gaps it is possible to estimate the sequences of quite ancient proteins (Hall 2006). The process of estimating and correcting for ancestral gaps was awkward and time-consuming, but it worked. Now MEGA has greatly simplified that process by estimating ancestral sequences that *include* the ancestral gaps, making it necessary only to remove those gaps to generate the ancestral sequence.

Using MEGA to Estimate Ancestral Sequences by Maximum Likelihood

Create the alignment

Use MEGA to download the coding sequences for the proteins of interest, as described in Chapter 3. Be sure to include an outgroup sequence that will allow the tree to be rooted. From the **Alignment Editor** export the alignment in FASTA format. Because ancestral sequence estimation is so sensitive to accurate alignment, use the GUIDANCE program (see Chapter 12) to align the sequences by codons and check the reliability with which each sequence aligns. If necessary, eliminate any poorly aligning sequences and re-align. The file `MinData.fas` includes the Gram negative sequences from `SmallData`, plus the Gram positive *G. vaginalis* sequence as an outgroup.

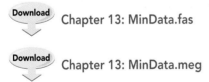

Download   Chapter 13: MinData.fas

Download   Chapter 13: MinData.meg

Construct the phylogeny

Open the FASTA alignment in MEGA and export it as a .meg file. Estimate a Maximum Likelihood tree as described in Chapter 9 with the following *essential* modification: in the **Analysis Preferences** window set **Gaps/Missing Data Treatment** to **Use all sites** as is shown in **Figure 13.1**. It is necessary to include all sites in those sequences in order to estimate the ancestral sequences.

Figure 13.1

Options Summary	
Option	Selection
Analysis	Phylogeny Reconstruction
Statistical Method	Maximum Likelihood
Phylogeny Test	
Test of Phylogeny	None
No. of Bootstrap Replications	Not Applicable
Substitution Model	
Substitutions Type	Nucleotide
Genetic Code Table	Not Applicable
Model/Method	General Time Reversible model
Rates and Patterns	
Rates among Sites	Gamma Distributed (G)
No of Discrete Gamma Categories	5
Data Subset to Use	
Gaps/Missing Data Treatment	Use all sites
Site Coverage Cutoff (%)	Complete deletion
Select Codon Positions	Use all sites / Partial deletion
Tree Inference Options	
ML Heuristic Method	Nearest-Neighbor-Interchange (NNI)

✓ Compute   ✗ Cancel   ? Help

Root the tree on the outgroup sequence that you included and save the tree as an `.mts` file.

Download

Chapter 13: MinData.mts

Examine the ancestral states at each site in the alignment

With no effort on your part, MEGA already has all the information needed to obtain the ancestral sequences at each of the internal node. To see that information, root the tree on the outgroup, then choose **Show All** from the **Ancestors** menu of the **Tree Explorer** window. Because it is easier to see if the tree is in the cladogram format, choose **Topology Only** from the **View** menu. The tree is now displayed with the estimated base at each node for the first site in the alignment (**Figure 13.2**).

Figure 13.2

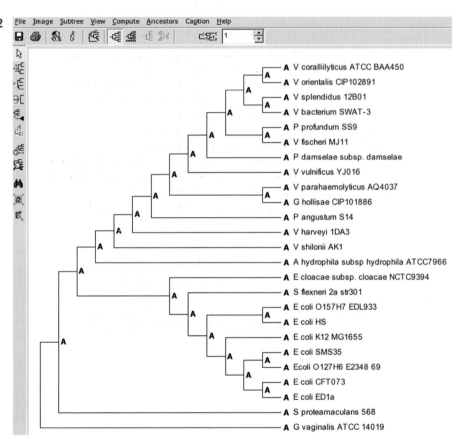

Clicking the up or down arrows (cursor arrow in **Figure 13.3**) or entering a number in the text box allows you to move through the sites in the alignment. Go to site 7 as shown in Figure 13.3.

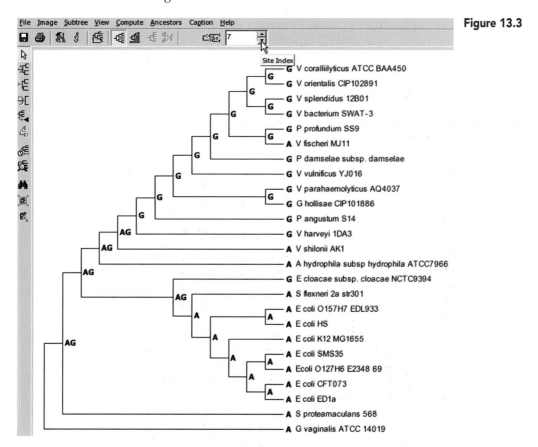

Figure 13.3

Notice in Figure 13.3 that some nodes show more than one possible base. To see the most probable base at each node, choose **Show most probable** from the **Ancestors** menu (**Figure 13.4**).

Figure 13.4

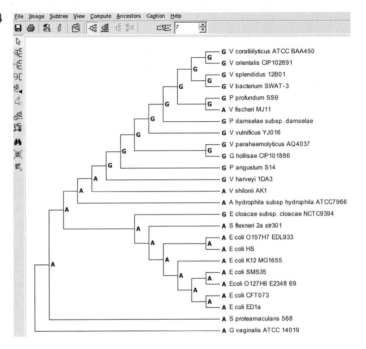

For some purposes (e.g., for seeing how key residues encoding active sites evolved), it is sufficient just to look at the states at a few sites in this fashion. However, if we are interested in the complete ancestral sequences, it is necessary to extract a bit more information

Estimate the ancestral sequence

Three steps are required to estimate the ancestral sequence:

1. Determine the numbering of each internal node.

2. Export a file that lists the probability of each base at each site for each of those nodes.

3. Extract the most probable DNA and protein sequences from the file using the Utility program ExtAncSeqMEGA (see Chapter 11). ExtAncSeqMEGA reports the ancestral sequence with and without ancestral gaps, along with the corresponding ancestral protein sequences. Importantly, it also reports accuracy scores for those ancestral sequences.

NUMBER THE INTERNAL NODES Turn off the display of bases at each node by choosing **None** from the **Ancestors** menu, then print the tree. (Mac and Linux users will need to save the tree image as a PDF file then print the PDF.)

Choose **Show All** from the **Ancestors** menu, then from the **Ancestors** menu choose **Export most probable sequences**. In the resulting **Select Output Format** window, change the **Output Option** to **Text editor (Display)** (**Figure 13.5**).

The results will be displayed in MEGA's **Text File Editor and Format Converter** window (**Figure 13.6**). From that window, save the file with an appropriate name.

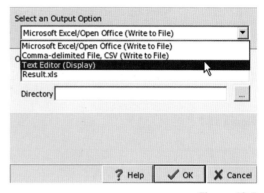

Chapter 13: MinData Most
Probable Sequences.txt

Figure 13.5

The file lists each node and the estimated sequence at that node. The external (leaf) nodes are identified by name and number and the internal nodes (boxed in Figure 13.6) by number. Notice that for nodes 33 and 34 the seventh character is "R," an ambiguity symbol that means "either A or G," as shown in Figure 13.3.

Figure 13.6

```
Most Probable Sequences
1:                                          234567891011121314151 61
1.  E coli K12 MG1655:                      ATGAGGATCATCGATAACTTAGA
2.  S flexneri 2a str301:                   ATGAGGATCATCGATAACTTAGA
3.  E coli O157H7 EDL933:                   ATGAGGATCATCGATAACTTAGA
4.  E cloacae subsp. cloacae NCTC9394:      ATGATCGTTCTTGAGAGCCTGGA
5.  P damselae subsp. damselae:             ATGATCGTACTTGATAACTTAGA
6.  V harveyi 1DA3:                         ATGATCGTGTTAGACAACTTAGA
7.  V shilonii AK1:                         ATGATCATTTAGAGAGCTTAGA
8.  V vulnificus YJ016:                      ATGATCGTGTTAGACAGCTTAGA
9.  V coralliilyticus ATCC BAA450:          ATGATCGTGTTAGAAAACTTAGC
10. V parahaemolyticus AQ4037:              ATGATCGTGTTAGACAGCTTAGA
11. V splendidus 12B01:                      ATGATCGTTTTAGACAACCTAGA
12. V orientalis CIP102891:                 ATGATCGTGTTAGAGAGCTTAGA
13. P profundum SS9:                        ATGGTTGTTTTAGAGAACCTCGA
14. G hollisae CIP101886:                   ATGATCGTGTTAGACAGCTTGGC
15. P angustum S14:                         ATGATCGTATTAGATAGCTTAGC
16. V bacterium SWAT-3:                     ATGATCGTTTTAGAGAACTTAGA
17. V fischeri MJ11:                        ATGATCATTTTAGACAATCTAGA
18. A hydrophila subsp hydrophila ATCC7966: ATGATCACCCTGGACAACCTG
19. S proteamaculans 568:                   ATGATTATTGTCGAATCATTAGC
20. G vaginalis ATCC 14019:                 ATGCGTATTTATAACGATGTTTC
21. E coli SMS35:                           ATGAGGATCATCGATAACTTAGA
22. E coli CFT073:                          ATGAGGATCATCGATAACTTAGA
23. Ecoli O127H6 E2348 69:                  ATGAGGATCATCGATAACTTAGA
24. E coli ED1a:                            ATGAGGATCATCGATAACTTAGA
25. E coli HS:                              ATGAGGATCATCGATAACTTAGA
26. (22 . 24):                             ATGAGGATCATCGATAACTTAGA
27. (28 . 26):                             ATGAGGATCATCGATAACTTAGA
28. (21 . 23):                             ATGAGGATCATCGATAACTTAGA
29. (3 . 25):                              ATGAGGATCATCGATAACTTAGA
30. (29 . 31):                             ATGAGGATCATCGATAACTTAGA
31. (1 . 27):                              ATGAGGATCATCGATAACTTAGA
32. (2 . 30):                              ATGAGGATCATCGATAACTTAGA
33. (4 . 32):                              ATGAYCRTYHTYGANAVCYTRGA
34. (36 . 33):                             ATGATCRTNHTNGANANACYTRGV
35. (34 . 19):                             ATGATYATYSTYGAVHNMYTAGH
36. (39 . 18):                             ATGATCRTNYTNGANAVCYTRGV
37. (10 . 14):                             ATGATCGTGTTAGACAGCTTAGM
38. (11 . 16):                             ATGATCGTYTTAGASAACYTAGA
```

Next to each taxon name on the printed tree, write that taxon's number, then number each internal node. Thus, for instance, node 26 is the parent of nodes 22 and 24, i.e., of `E coli CFT073` and `E coli ED1A`. There are a total of 48 nodes (not all are shown in Figure 13.6), and each can be labeled by working from the leaves toward the root. **Figure 13.7** shows this for a Maximum Likelihood tree on which the nodes are manually numbered in red.

Figure 13.7

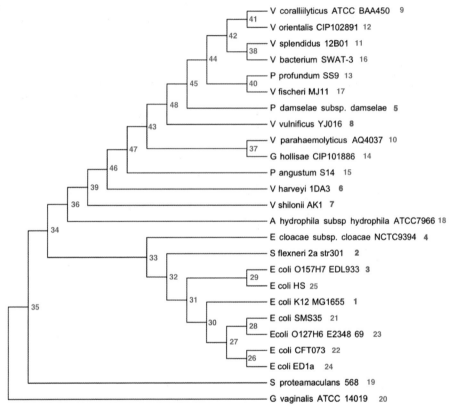

EXPORT THE DETAILED TEXT EXPORT FILE In the **Ancestors** menu be sure that **Show All** is ticked, then choose **Detailed Text Export** and give the file a sensible name. I like to use the extension `.anc` to remind me that this is a file of ancestral sequences, e.g., `MyFile.anc`.

 Chapter 13: MinData.anc

EXTRACT THE ANCESTRAL SEQUENCES FROM THE .ANC FILE Use the utility program ExtAncSeqMEGA to extract the ancestral DNA and protein sequences. In Terminal (Mac or Unix) or Command Prompt (Windows) navigate to the folder that contains the `.anc` file. Enter `ExtAncSeqMEGA MyFile.anc`. The program will create a folder named **MyFile.anc Output files** into which it will write one file for each internal node on the tree. For the example in Figure 13.7, the files will be named Node 26 through Node 48.

The top section of each file lists information about the ancestral DNA sequence at a node. For each site in the alignment it shows the probabilities of an A, C, G, or T at that site, followed by the most probable base (MPB) at that site and the probability of the MPB. If that site is most likely a gap, it shows 0 for each probability and a gap character (–) as the MPB. **Figure 13.8** shows this information for node 33.

Figure 13.8

```
Node 33
Ancestral DNA sequence:
Probabilities of bases and MPB (most probable base) at each site.
Site     A         C         G         T         MPB       Probability
1        1         0         0         0         A         1
2        0         0         0         1         T         1
3        0         0         1         0         G         1
4        1         0         0         0         A         1
5        0         0.07      0.01      0.92      T         0.92
6        0         0.96      0.02      0.01      C         0.96
7        0.63      0.01      0.36      0         A         0.63
8        0         0         0         1         T         1
9        0.05      0.39      0.03      0.53      T         0.53
10       0.14      0.75      0.02      0.09      C         0.75
11       0         0         0         1         T         1
12       0.03      0.61      0.01      0.35      C         0.61
13       0         0         1         0         G         1
14       1         0         0         0         A         1
15       0.14      0.3       0.31      0.25      G         0.31
16       1         0         0         0         A         1
17       0.34      0.06      0.58      0.02      G         0.58
18       0         1         0         0         C         1
```

Below the list of probabilities, the most probable ancestral DNA sequence for that node is shown with and without gaps (**Figure 13.9**). An accuracy score for the sequence without gaps is shown. The accuracy score is the mean of the probabilities of all the bases. An accuracy score of 0.8594 can be interpreted as meaning that, on average, any base has a 86% chance of being correct.

Figure 13.9

```
The most probable ancestral DNA sequence is:
ATGATCATTCTCGAGAGCTTAGAACAGTTTCAACAGATTTACCGCACCGGTAGGAAATGGCAGCGCTGCGTAGAAGCCATTGACAATAT

The most probable DNA sequence without gaps is:
ATGATCATTCTCGAGAGCTTAGAACAGTTTCAACAGATTTACCGCACCGGTAGGAAATGGCAGCGCTGCGTAGAAGCCATTGACAATAT

The accuracy score for that sequence is 0.859373601789709

The accuracy score is defined as the average probability of
the most probable bases.  It is NOT the probability that the
most probable sequence is correct.  That probability is 7.82298357693024e-37.
The natural log of that probability is -83.1385824274104.
```

The probability of the entire sequence being correct is the product of the probabilities of the MPB and is a very small number, e.g. 7.8×10^{-37}. Don't be dismayed by that small value. The log of that probability is –83.14, which is considerably higher than the lnL of the tree itself (–5982.42).

If the sequence is possibly a coding sequence—i.e., if its length is a multiple of three—the next section lists information about the encoded ancestral protein sequence (**Figure 13.10**).

Figure 13.10

```
Ancestral Protein Sequence:
Column headings:
MPAA = most probable amino acid, pMPAA = probability of MPAA
MPAA2 = second most probable amino acid, pMPAA2 = probability of MPAA2
Ratio = pMPAA/pMPAA2,  Unreliable means Ratio <1.5

MPAA     pMPAA      MPAA2     pMPAA2     Ratio
M        1          A         0          9999
I        0.8924     T         0.0693     12.8773448773449
I        0.6111     V         0.36       1.6975
L        0.7536     I         0.1386     5.43722943722944
D        0.55       E         0.45       1.22222222222222     Unreliable
S        0.58       N         0.34       1.70588235294118
L        0.9832     F         0.0068     144.588235294118
E        0.912      D         0.038      24
Q        1          A         0          9999
F        1          A         0          9999
```

For each site in the ancestral protein, the most probable amino acid (MPAA) and its probability are shown, followed by the second most probably amino acid and its probability. Column 5 shows the ratio of the probability of the MPAA to the second MPAA. When the probability of the second MPAA is 0 the ratio is infinity, indicated by 9999. By default, when that ratio is <1.5 the site is flagged as "Unreliable" in column 6. That cutoff is purely arbitrary and serves only to alert you that you might not want to take the amino acid at that position too seriously. You can change the cutoff ratio by adding a new cutoff as an argument when you run ExtAncSeqMEGA. For example, enter `ExtAncSeqMEGA MyFile.anc 2.0` to set the cutoff to the more stringent 2.0 instead of the default 1.5.

Finally, the sequences of the ancestral protein with and without gaps are shown, as are the accuracy scores for the ungapped sequence and the probability and log probability that the sequence is correct (**Figure 13.11**).

Figure 13.11

```
The most probable ancestral protein sequence is:
MIILDSLEQFQQIYRTGRKWQRCVEAIDNIDNIQPGVAHSIGDSLAYRVENDAG-TDALFVGHRRYFEVHYYLQGQQKIEYAAKDALQVVE(

The most probable ancestral protein sequence without gaps is:
MIILDSLEQFQQIYRTGRKWQRCVEAIDNIDNIQPGVAHSIGDSLAYRVENDAGTDALFVGHRRYFEVHYYLQGQQKIEYAAKDALQVVEC'

The accuracy score for that sequence is 0.845800412162162

The accuracy score is defined as the average probability of
the most probable amino acids.  It is NOT the probability that the
most probable sequence is correct.  That probability is 1.60302391656126e-14
The natural log of that probability is -31.7642995085245.
```

If the ancestral DNA sequence has gaps *within* a codon (i.e., a frame shift would result), an error message reports that the codon is bad and that calculation of the ancestral protein sequence is terminated. That should not happen if the sequences are aligned by codons.

Calculating the ancestral protein sequence and amino acid probabilities

The ancestral protein sequence is not calculated simply by translating the ancestral DNA sequence. Instead, the probabilities of each of the possible 64 codons are calculated, and the probabilities of each amino acid are determined by summing the probabilities of all codons that encode that amino acid. The ancestral protein sequence is composed of the most likely amino acids calculated in this manner.

How Accurate are the Estimated Ancestral Sequences?

Hall (2006) studied the accuracy of ancestral sequences estimated by Bayesian Inference by simulating evolution over several different trees. In that situation the true sequences of the internal nodes were known and could be compared with the estimated sequences. As might be expected, accuracy decreased as the ancestral nodes were closer to the root. The accuracies of estimated ancestral protein sequences were significantly higher than those of estimated ancestral DNA sequences because many of the errors in DNA sequences did not affect the encoded amino acid. The study also compared the actual accuracies with the accuracy scores (the average probabilities of the bases or amino acids).

For Bayesian Inference by MrBayes, the accuracies of the ancestral DNA sequences (mean ± s.e.) were 0.899 ± 0.015, and accuracies of the protein sequences were 0.969 ± 0.007. That can be compared with Maximum Likelihood estimates by MEGA, where the accuracies of ancestral DNA sequences were 0.904 ± 0.014 and those of protein sequences were 0.963 ± 0.09. The accuracies of the two methods are indistinguishable.

MEGA's accuracy score overestimates accuracy of ancestral DNA sequences by 1.4 ± 0.5%, while MrBayes overestimates accuracy by 2.7 ± 0.8%. MEGA underestimates accuracy of ancestral protein sequences by 0.4 ± 0.1%, while MrBayes underestimates accuracy by 0.8 ± 0.2%. For both methods, the accuracy score is a good reflection of the overall accuracy of estimated ancestral sequence.

CHAPTER 14

Detecting Adaptive Evolution

Most amino acid substitutions are deleterious (i.e., they decrease the function of the protein) and are eventually eliminated from the population via **negative** or **purifying** selection. Some amino acid substitutions are neutral, and pure chance determines whether they will be fixed into the population (i.e., whether the allele carrying the replacement will become the most prevalent in the population). In rare instances, an amino acid substitution is beneficial and is fixed into the population by **positive** or **diversifying** selection. At the nucleotide level, some base substitutions cause amino acid replacements and others are silent (i.e., they do not change the amino acid specified). Silent mutations are neutral or nearly so.

When homologous DNA sequences are compared, it is almost always the case that silent mutations outnumber replacement mutations. This is attributed to purifying selection eliminating most amino acid replacement mutations. If altered environmental conditions or new demands favor modifications of a protein's function, then amino acid substitutions that enhance that modification will be subject to positive selection. Because silent mutations tend to accumulate roughly linearly over time, during periods of positive selection amino acid replacement mutations will tend to exceed silent mutations. With this in mind, evolutionary biologists interpret an excess of replacement mutations as evidence of positive selection (i.e., something important was happening to the function of a protein).

Most mutations at the first position in a codon result in amino acid replacements; all mutations at second positions replace the specified amino acid. Because of the redundancy of the genetic code, however, many third-position mutations are silent. To identify evidence of either positive or purifying selection, molecular evolutionists normalize the number of replacement and silent mutations to the number of replacement and silent sites in the gene. The number of replacement mutations per replacement site is called dN (sometimes symbolized β), and the number of silent mutations per silent site is called dS (sometimes α). The dN/dS ratio (sometimes ω) is often used as a measure of the operation of selection. When sequences are compared, a dN/dS ratio less than 1 is taken as evidence of purifying selection; a ratio of 1.0 is taken as evidence that mutations were generally neutral; and a dN/dS ratio greater than 1 is taken as evidence of positive selection. The intensity of selection is reflected in the magnitude of the dN/dS ratio. In keeping with this expectation, comparisons of pseudogenes—whose products contribute nothing to fitness—usually show dN/dS ratios very close to one.

There are a number of levels at which we can pose questions about the operation of selection. Consider a set of homologous genes whose coding sequences have been aligned. We can ask:

- *Since the time those genes diverged from a common ancestor, have the genes been undergoing neutral, purifying, or positive selection?* To answer that we can estimate the overall dN/dS ratio.

- *Has purifying, positive, or neutral evolution dominated since any two particular sequences diverged from their common ancestor?* To answer that we can estimate the dN/dS ratio by comparing just those two sequences.

- *When did selection occur?* To answer that we can estimate the dN/dS ratio along a particular branch of the tree.

- *Where in the gene did selection occur?* To answer that we can estimate dN/dS at particular codons in the alignment.

It may be important to consider the issue of selection at each of those levels. Imagine that phylogenetic analysis shows that within a set of glycosidase genes, those that act on α substrates diverged from a set that act on β substrates but that they certainly share a common ancestor. It may well be the case that overall those genes have been subjected to strong purifying selection, but that during the period immediately following the divergence of the two groups there was strong selection favoring those substitutions that increased specificity for their respective substrates. Furthermore, if we look carefully, it may be that those favored substitutions occurred at only a few sites within the gene—those that involved the active site's binding specificity.

Several approaches for estimating dN and dS are discussed in detail in Chapter 4 of *Molecular Evolution and Phylogenetics* (Nei and Kumar 2000). I am concerned here with two of these approaches: (1) an Evolutionary Pathway method (the Nei-Gojobori method as implemented by MEGA); and (2) a Maximum Likelihood method implemented by the codeml program of the PAML package (Yang 1997). Since the entire basis of both these analyses is the comparison of codons, it is essential that the sequences be aligned according to the protein sequence alignment described in Chapter 4.

This chapter will use the data set HIVenvSweden, a set of sequences from the V3 region of the *env* gene of the HIV virus. That region is thought to evolve rapidly in response to anti-HIV drugs and might be expected to be undergoing positive selection (Yang et al. 2000). The sequences in this example align very reliably without gaps (their GUIDANCE sequences scores are all ≥0.999).

Effect of Alignment Accuracy on Detecting Adaptive Evolution

Fletcher and Yang (2010) used simulations to study the effects of alignment accuracy on detecting adaptive evolution and found that alignment accuracy strongly affects the frequency of false positives in detecting branches that are under positive selection. They found that PRANK outperformed MUSCLE and MAFFT, both of which outperformed ClustalW. Errors were reduced by eliminating columns that contained gaps.

That study suggests that it is very advisable to use GUIDANCE (see Chapter 12) to align sequences when studies of adaptive evolution are anticipated. Because GUIDANCE permits eliminating those columns that align unreliably, it provides a more sophisticated approach than does simply eliminating all columns with gaps.

Using MEGA to Detect Adaptive Evolution

Detecting overall selection

In MEGA's main window, open HIVenvSweden.meg

Download Chapter 14: HIVenvSweden.meg

To test the hypothesis that positive selection was operating as these sequences diverged, choose **Codon-Based Z-test of selection** from the **Selection** menu. In the **Analysis Preferences** window be sure **Gaps/Missing Data** is set to **Pairwise Deletion**; **Analysis Scope** is set to **Overall Average**; **Variance Estimation Method** is set to **Bootstrap** with 100 replications; and set **Test Hypothesis (HA: alternative)** to **Positive Selection (HA: dN>dS)**. Set the **Substitution Type** by choosing **Syn-Nonsynonymous** and the choosing **Nei-Gojobori method (proportion)** (**Figure 14.1**). Click the **Compute** button.

Figure 14.1

Option	Selection
Analysis	Z-test of Selection
Scope	Overall Average
Test Hypothesis (HA: alternative)	Positive selection (HA: dN > dS)
Estimate Variance	
Variance Estimation Method	Bootstrap method
No. of Bootstrap Replications	100
Substitution Model	
Substitutions Type	Syn-Nonsynonymous
Model/Method	Nei-Gojobori method (Proportion)
Fixed Transition/Transversion Ratio	Not Applicable
Data Subset to Use	
Gaps/Missing Data Treatment	Pairwise deletion
Site Coverage Cutoff (%)	Not Applicable

Options Summary

✓ Compute   ✗ Cancel   ? Help

The resulting window shows that the overall probability that dN exceeds dS by the observed amount by chance alone is 0.085 (**Figure 14.2**). Since we usually consider p values >0.05 not to be significant, we conclude that we cannot reject the hypothesis of strictly neutral evolution in favor of the alternate hypothesis of positive evolution. If we test purifying selection by changing **Test Hypothesis** to **Purifying Selection**, we get $p = 1.00$, so again we cannot reject the hypothesis of strict neutrality. Similar testing of the hypothesis of neutrality (dN = dS) gives $p = 0.115$, so we cannot reject the hypothesis of strict neutrality either. We simply cannot draw any significant conclusion about selection averaged over all the sequences.

Figure 14.2

Detecting selection between pairs

If we change the **Analysis Scope** to **In Sequence Pairs** and test the hypothesis of **Positive Selection** (**Figure 14.3**), we see a matrix in the results window (**Figure 14.4**). The test statistic, dN – dS, referred to as the Z value, is shown in blue type above the diagonal and p is below the diagonal.

Figure 14.3

Option	Selection
Analysis	Z-test of Selection
Scope	In Sequence Pairs
Test Hypothesis (HA: alternative)	Positive selection (HA: dN > dS)
Estimate Variance	
Variance Estimation Method	Bootstrap method
No. of Bootstrap Replications	100
Substitution Model	
Substitutions Type	Syn-Nonsynonymous
Model/Method	Nei-Gojobori method (Proportion)
Fixed Transition/Transversion Ratio	Not Applicable
Data Subset to Use	
Gaps/Missing Data Treatment	Pairwise deletion
Site Coverage Cutoff (%)	Not Applicable

$\checkmark$ Compute $\times$ Cancel   ? Help

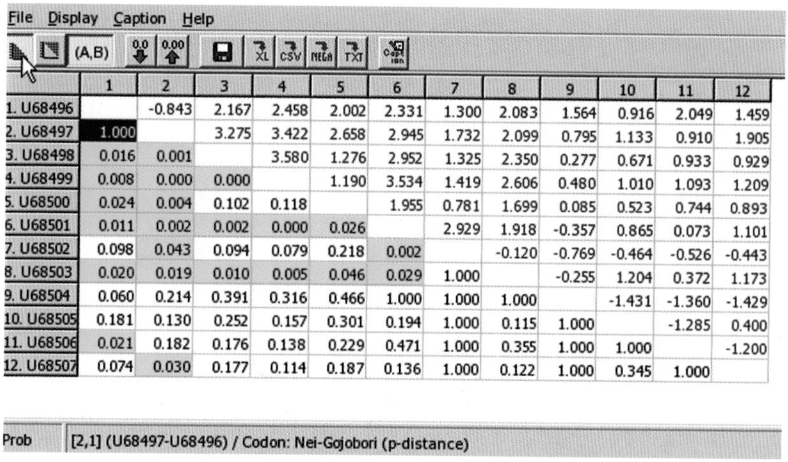

Figure 14.4

File Display Caption Help

	1	2	3	4	5	6	7	8	9	10	11	12
1. U68496		-0.843	2.167	2.458	2.002	2.331	1.300	2.083	1.564	0.916	2.049	1.459
2. U68497	1.000		3.275	3.422	2.658	2.945	1.732	2.099	0.795	1.133	0.910	1.905
3. U68498	0.016	0.001		3.580	1.276	2.952	1.325	2.350	0.277	0.671	0.933	0.929
4. U68499	0.008	0.000	0.000		1.190	3.534	1.419	2.606	0.480	1.010	1.093	1.209
5. U68500	0.024	0.004	0.102	0.118		1.955	0.781	1.699	0.085	0.523	0.744	0.893
6. U68501	0.011	0.002	0.002	0.000	0.026		2.929	1.918	-0.357	0.865	0.073	1.101
7. U68502	0.098	0.043	0.094	0.079	0.218	0.002		-0.120	-0.769	-0.464	-0.526	-0.443
8. U68503	0.020	0.019	0.010	0.005	0.046	0.029	1.000		-0.255	1.204	0.372	1.173
9. U68504	0.060	0.214	0.391	0.316	0.466	1.000	1.000	1.000		-1.431	-1.360	-1.429
10. U68505	0.181	0.130	0.252	0.157	0.301	0.194	1.000	0.115	1.000		-1.285	0.400
11. U68506	0.021	0.182	0.176	0.138	0.229	0.471	1.000	0.355	1.000	1.000		-1.200
12. U68507	0.074	0.030	0.177	0.114	0.187	0.136	1.000	0.122	1.000	0.345	1.000	

Prob   [2,1] (U68497-U68496) / Codon: Nei-Gojobori (p-distance)

When dN – dS is positive, positive selection has operated since the two sequences diverged. The value of p represents the probability of that great a difference or more arising by chance alone; thus when $p < 0.05$ there has been significant positive selection (highlighted). In column 1, row 2 has a p value of 1.0, because the test for significance of the Z value is one-tailed. Selecting the p value of 1.0, then clicking the triangle button (arrow cursor in Figure 14.4) automatically selects the corresponding Z value of –0.843 (**Figure 14.5**). The negative value means that dS was actually greater than dN, which is an indication of possible purifying selection.

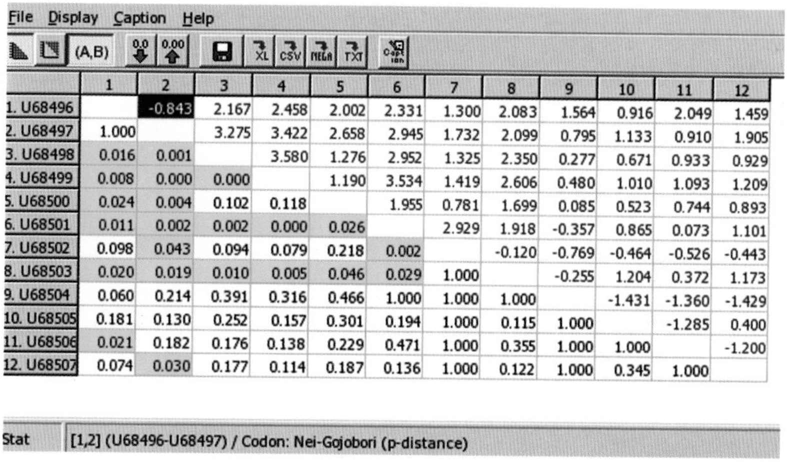

Figure 14.5

File Display Caption Help

	1	2	3	4	5	6	7	8	9	10	11	12
1. U68496		-0.843	2.167	2.458	2.002	2.331	1.300	2.083	1.564	0.916	2.049	1.459
2. U68497	1.000		3.275	3.422	2.658	2.945	1.732	2.099	0.795	1.133	0.910	1.905
3. U68498	0.016	0.001		3.580	1.276	2.952	1.325	2.350	0.277	0.671	0.933	0.929
4. U68499	0.008	0.000	0.000		1.190	3.534	1.419	2.606	0.480	1.010	1.093	1.209
5. U68500	0.024	0.004	0.102	0.118		1.955	0.781	1.699	0.085	0.523	0.744	0.893
6. U68501	0.011	0.002	0.002	0.000	0.026		2.929	1.918	-0.357	0.865	0.073	1.101
7. U68502	0.098	0.043	0.094	0.079	0.218	0.002		-0.120	-0.769	-0.464	-0.526	-0.443
8. U68503	0.020	0.019	0.010	0.005	0.046	0.029	1.000		-0.255	1.204	0.372	1.173
9. U68504	0.060	0.214	0.391	0.316	0.466	1.000	1.000	1.000		-1.431	-1.360	-1.429
10. U68505	0.181	0.130	0.252	0.157	0.301	0.194	1.000	0.115	1.000		-1.285	0.400
11. U68506	0.021	0.182	0.176	0.138	0.229	0.471	1.000	0.355	1.000	1.000		-1.200
12. U68507	0.074	0.030	0.177	0.114	0.187	0.136	1.000	0.122	1.000	0.345	1.000	

Stat   [1,2] (U68496-U68497) / Codon: Nei-Gojobori (p-distance)

If we repeat the analysis, but change **Test Hypothesis** to **Purifying Selection**, we find no evidence for significant purifying selection. We can conclude that the evolution of these genes has been under strong positive selection during some, but not all of its history.

Finding the region of the gene that has been subject to positive selection

Choose **Estimate Selection for each Codon (HyPhy)** from the **Selection** menu in MEGA's main window. The **Analysis Preferences** window has two setting that require attention (**Figure 14.6**).

Figure 14.6

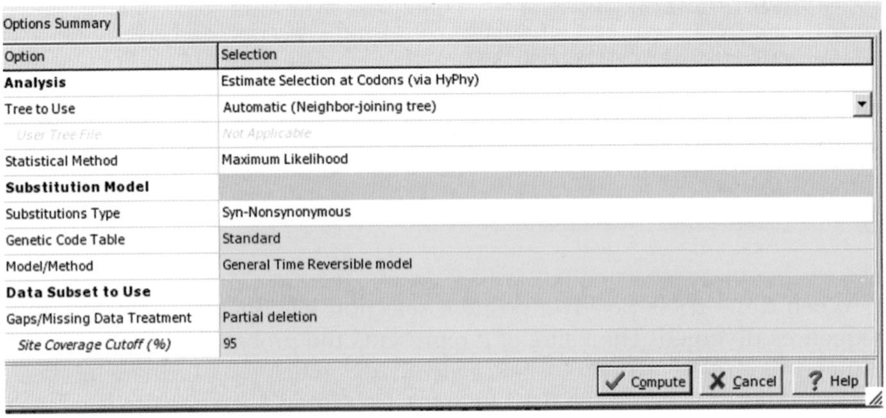

Tree to Use has as its default an NJ tree. The alternative is **Specify a tree from a file**. In general I am more comfortable with an ML than an NJ tree, so I estimated an ML tree under the GTR+G+I model and exported it as a Newick tree, then clicked in the **User Tree File** row and navigated to the Newick file (**Figure 14.7**). (The other choice is **Model/Method**. Use MEGA's **Find best DNA/Protein Models (ML)** choice from the **Models** menu to choose an appropriate model.)

Figure 14.7

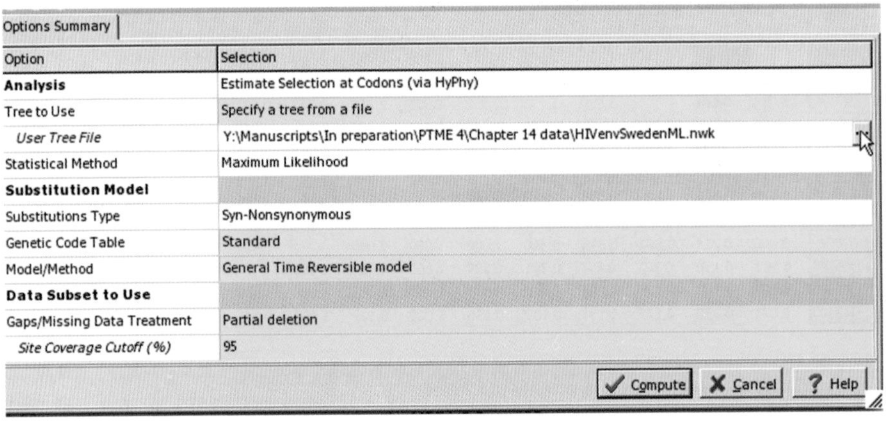

When you click the **Compute** button, a dialog box appears from which you can choose how to view the results (**Figure 14.8**). The default is a Microsoft Excel file with the default name `Result.xls`. If you prefer that option, you should at least change the file name to something meaningful (but keep the .xls extension) and specify an appropriate directory.

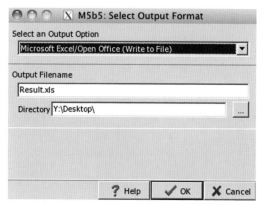

Figure 14.8

The alternatives are to save the results as a **Comma-delimited CSV** file that can be imported into most spreadsheet programs, or not to save it at all but to display it in MEGA's own text editor (**Figure 14.9**).

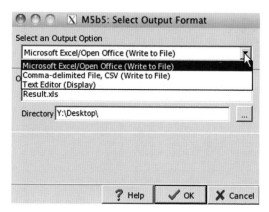

Figure 14.9

Figure 14.10 shows a portion of the Excel file. Column B lists the codons that are in the first sequence in the alignment. The columns of interest are columns I and J. Column I shows the test statistic dN – dS. Codons for which that value is positive show evidence of positive selection, those for which it is negative evidence of purifying selection. Column J shows the probability that positive values are significantly different from 1.0 (highlighted). Notice that all of the highlighted values are zero, and that in each case both dN (column H) and dS (column G) are both zero. Those highlighted cells are not, after all, significant.

Figure 14.10

	A	B	C	D	E	F	G	H	I	J	K
1	Codon#	Triplet	Syn (s)	Nonsyn (n)	Syn sites (S)	Nonsyn sites (N)	dS	dN	dN-dS	P-value	Normalized dN-dS
2	1	GTA	0	4	0.813203	2.1868	0	1.8292	1.82916	0.2823	3.52598
3	2	GTA	0	0	1	2	0	0	0	0	0
4	3	ATT	0	0	0.866014	2.13399	0	0	0	0	0
5	4	AGA	1	0	0.86226	1.98018	1.1597	0	-1.1597	1	-2.23558
6	5	TCT	0	0	1	2	0	0	0	0	0
7	6	GAA	0	0	0.560309	2.33999	0	0	0	0	0
8	7	AAC	0	0	0.359797	2.6402	0	0	0	0	0
9	8	TTC	0	0	0.359797	2.6402	0	0	0	0	0
10	9	TCG	2	1	1	1.93358	2	0.5172	-1.4828	0.9604	-2.85837
11	10	AAC	0	1	0.359797	2.6402	0	0.3788	0.37876	0.8801	0.730116
12	11	AAT	0	0	0.555579	2.44442	0	0	0	0	0
13	12	GCT	0	0	1	2	0	0	0	0	0
14	13	AAA	2	0	0.591411	2.25533	3.3818	0	-3.3818	1	-6.51883
15	14	ACC	0	0	1	2	0	0	0	0	0
16	15	ATA	0	0	0.439691	2.56031	0	0	0	0	0
17	16	ATA	0	0	0.439691	2.56031	0	0	0	0	0
18	17	GTA	0	0	1	2	0	0	0	0	0
19	18	CAG	0	1	0.834965	1.81717	0	0.5503	0.55031	0.6852	1.0608
20	19	CTA	0	0	1.3598	1.6402	0	0	0	0	0
21	20	AAT	2	1	0.551068	2.43788	3.6293	0.4102	-3.2191	0.9937	-6.20535
22	21	AAA	0	3	0.54753	2.31462	0	1.2961	1.29611	0.5289	2.49845
23	22	TCT	0	2	1	2	0	1	1	0.4444	1.92765
24	23	GTA	0	0	1	2	0	0	0	0	0
25	24	GAA	0	3	0.573869	2.33753	0	1.2834	1.28341	0.5176	2.47396
26	25	ATT	0	0	0.866014	2.13399	0	0	0	0	0
27	26	AAT	0	5	0.628553	2.36012	0	2.1185	2.11854	0.3071	4.08381
28	27	TGT	0	0	0.555579	2.13399	0	0	0	0	0
29	28	ACA	0	8	0.699587	2.23697	0	3.5763	3.57626	0.1134	6.89379
30	29	AGA	0	0	0.846745	2	0	0	0	0	0
31	30	CCC	0	0	1	2	0	0	0	0	0
32	31	AAC	0	3	0.367286	2.63271	0	1.1395	1.13951	0.6758	2.19658
33	32	AAC	0	0	0.359797	2.6402	0	0	0	0	0
34	33	AAT	0	0	0.555579	2.44442	0	0	0	0	0
35	34	ACA	0	1	1	2	0	0.5	0.5	0.6667	0.963827
36	35	AGA	0	1	0.833659	2.01309	0	0.4968	0.49675	0.7072	0.957561
37	36	AGA	0	3	0.591981	2.26177	0	1.3264	1.3264	0.4978	2.55683
38	37	AGT	0	1	0.568882	2.42412	0	0.4125	0.41252	0.8099	0.795198
39	38	ATA	0	1	0.502815	2.49718	0	0.4005	0.40045	0.8324	0.771931
40	39	CAT	0	3	0.583903	2.4161	0	1.2417	1.24167	0.5224	2.39351
41	40	TTT	2	4	0.448337	2.54121	4.4609	1.5741	-2.8869	0.9527	-5.5649
42	41	GGA	0	1	0.996808	1.90237	0	0.5257	0.52566	0.6562	1.01329
43	42	CCA	0	0	1	2	0	0	0	0	0

Indeed, there are *no* codons for which there is significant evidence of positive selection. Keep in mind that those values are averaged over the entire tree. If some codons were under positive selection, but only over a few branches, their overall selection might not be detectable by this test.

Using Codeml to Detect Adaptive Evolution

The program codeml, which is part of the PAML package developed by Ziheng Yang, employs a Maximum Likelihood approach to estimate the dN/dS ratio for each branch in a phylogeny. You can download PAML from http://abacus. gene.ucl.ac.uk/software/paml.html. The site provides complete instructions for downloading and installing PAML; it is important to read those instructions. PAML includes a documentation folder named `doc`. Read through that documentation PDF! I will not attempt to re-create it here. PAML is actually a suite of programs; we will only be concerned with codeml.

Installation

WINDOWS Use a utility such as WinZip or Stuffit for Windows to unpack the PAML archive. The Windows executable files are in a folder named `bin` within the PAML folder. Move or copy `codeml.exe` to the `bin` folder that you already set up (see Chapter 11).

MACINTOSH OSX If you are using an Intel-based Mac and OSx10.5 or above, the downloaded file will automatically unpack to give you a PAML folder. The executables (programs) are in the `bin` folder within the PAML folder. To install codeml, open a Terminal window (see Chapter 11) and navigate to the bin folder within the PAML folder. You need to change the permissions so that everyone can run codeml. To do so enter `chmod 711 codeml` To install codeml, enter `sudo cp codeml /usr/local/bin/`, which copies codeml to the `usr/local/bin` folder and thus allows you to use it from within any folder. Close the Terminal window.

UNIX/LINUX Download any of the archives and unpack it. Unix/Linux users must compile codeml from the Source Code folder. Make the `src` folder within the PAML folder the current folder. Type `make -f makefile.unix` to compile all of the PAML programs including codeml, then type `rm *.o` (the letter "oh," not zero) to remove all of the object files. Finally, copy codeml to the directory you use for executables.

The files you need to run codeml

PAML programs are not interactive; instead they employ a **control file** that contains the instructions to the program and two input files that provide the data. One of the two input files is a sequence alignment in PHYLIP format.

THE ALIGNMENT FILE The first input file is a sequence alignment in the PHYLIP format (see Appendix I). PHYLIP, which is required by PAML programs, is unusual in requiring two spaces between the sequence name and the sequence, and the sequence name is limited to 30 characters. To make an alignment file in the PAML PHYLIP format, export the alignment from MEGA in

FASTA format, then use FastaConvert (see Chapter 11) to convert it to the PAML PHYLIP format by using the −x argument. The file will have the extension .pam.

If the alignment file was obtained elsewhere and is in the Interleaved format, you will need to put an I on the first line, separated by a space from the number of characters per sequence (e.g., 10 960 I).

THE TREE FILE The second input file is a tree, in Newick format. Use MEGA to make a Maximum Likelihood tree, then from the Tree viewer export the tree in Newick format.

THE CONTROL FILE The codeml control file serves the same function as a MrBayes block: it contains all of the commands to run the program. You can modify the example control file, codeml.ctl to suit your needs (**Figure 14.11**).

Figure 14.11

```
1       seqfile = myfile.txt
2      treefile = mytree.nwk
3       outfile = myoutput_0.out
4
5         noisy = 5    * 0,1,2,3,9: how much rubbish on the screen
6       verbose = 1    * 1: detailed output, 0: concise output
7       runmode = 0    * 0: user tree;  1: semi-automatic;  2: automatic
8                      * 3: StepwiseAddition; (4,5):PerturbationNNI
9
10      seqtype = 1    * 1:codons; 2:AAs; 3:codons-->AAs
11    CodonFreq = 2    * 0:1/61 each, 1:F1X4, 2:F3X4, 3:codon table
12        clock = 0    * 0: no clock, unrooted tree, 1: clock, rooted tree
13        model = 0
14                     * models for codons:
15                       * 0:one, 1:b, 2:2 or more dN/dS ratios for branches
16
17      NSsites = 0    * dN/dS among sites. 0:no variation, 1:neutral, 2:positive
18        icode = 0    * 0:standard genetic code; 1:mammalian mt; 2-10:see below
19
20    fix_kappa = 0    * 1: kappa fixed, 0: kappa to be estimated
21        kappa = 2    * initial or fixed kappa
22    fix_omega = 0    * 1: omega or omega_1 fixed, 0: estimate
23        omega = 2    * initial or fixed omega, for codons or codon-transltd AAs
24
25    fix_alpha = 1    * 0: estimate gamma shape parameter; 1: fix it at alpha
26        alpha = .0   * initial or fixed alpha, 0:infinity (constant rate)
27       Malpha = 0    * different alphas for genes
28        ncatG = 4    * # of categories in the dG or AdG models of rates
29
30        getSE = 0    * 0: don't want them, 1: want S.E.s of estimates
31  RateAncestor = 0   * (1/0): rates (alpha>0) or ancestral states (alpha=0)
32       method = 0    * 0: simultaneous; 1: one branch at a time
33
```

In the Control files, everything that follows an asterisk is a comment. Note that there *must* be a space on each side of an equals sign.

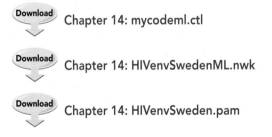

Download Chapter 14: mycodeml.ctl

Download Chapter 14: HIVenvSwedenML.nwk

Download Chapter 14: HIVenvSweden.pam

Only four modifications to the example Control file are needed. On the first three lines, change the names of the **seqfile**, the **treefile**, and the **outfile** to correspond to your input sequence and tree files and to any output file name you wish. File names are case-sensitive and must not include any spaces. Notice that I named the **outfile myoutput_0.out** to indicate that the output is from Model 0.

Finally, on the thirteenth line you must specify a model. If **model = 0**, then codeml calculates a single dN/dS ratio averaged over the entire tree (i.e., each branch has the same ratio). If **model = 1**, then codeml calculates an independent dN/dS ratio for each branch of the tree. We are most interested in model = 1, but we will actually run codeml twice, once with **model = 0** and once with **model = 1**. You might save the modified control file with a meaningful name such as **HIVenvSweden_0.ctl**. Then make a copy of the control file, save it under another name such as **HIVenvSweden_1.ctl**, change the thirteenth line to **model = 1** and the output name to **HIVenvSweden _1.out**.

 Chapter 14: HIVenvSweden_0.ctl

 Chapter 14: HIVenvSweden_1.ctl

The comments after each command in the control file describe the functions of that command and some alternative settings. Comments are ignored by PAML and are only there to remind you, the user, about the command. The commands are discussed in the **pamlDoc.pdf** documentation file in the **Doc** folder within the PAML folder that you downloaded. Many of the commands are the same as for the baseml program and are discussed in the part of the documentation devoted to baseml; the others are discussed in the codeml section.

Other than changing the **model** command (line 13 in Figure 14.11), you can leave the settings as shown in Figure 14.11. Later we will also modify the **fix_omega** command (line 22 in Figure 14.11).

Questions that underlie the models

Suppose that, over the course of evolution represented by the phylogeny, this gene had been under a constant level of selection. Such a scenario would mean that dN/dS would be about the same on all branches, but some branches might have dN/dS ratios >1. The first question we need to ask when we see variation in the dN/dS ratio among branches is, "Is that variation more than we would expect to see by chance alone when selection remains constant?" In codeml, Model 1 calculates the individual dN/dS ratio for each branch, while Model 0 calculates a single average dN/dS ratio, which is then assigned to each branch. We can compare the log likelihoods of the two models to see if the variation in dN/dS shown by Model 1 can be attributable to chance alone. If lnL (log likelihood) of Model 1 is significantly greater than that of Model 0 we can conclude that selection intensity and direction was not constant, and that the different dN/dS ratios are probably meaningful.

We identify those branches in which the dN/dS ratio is of special interest—perhaps those with dN/dS greater than 1.0, indicating positive or diversifying selection. We will want to know whether those dN/dS ratios are *significantly* greater than 1.0. To make that determination we can again compare two models. We first mark the branch of interest in the tree file (see below). We then run Model 2 under two conditions:

- In condition 2A, dN/dS of the branch of interest is calculated and the average dN/dS ratio of the remaining branches is assigned to those branches
- In condition 2B, the dN/dS ratio of the branch of interest is set to 1.0, and the average for the other branches is assigned to those branches.

If lnL of Model 2 under condition A is significantly greater than lnL of Model 2 under condition B, then the calculated dN/dS ratio is significantly greater than 1 and we can be confident that the gene was under significant positive selection along that branch.

Run codeml

To run codeml, Windows users should open Command Prompt; Macintosh OSX users should open Terminal (see Chapter 11). Use the `cd` command to navigate to the folder that contains the sequence file, the tree file, and the control file. For the example below I have used the `HIVenvSweden` sequence and tree files.

Be sure that the line endings are set correctly for your platform! Type `codeml` followed by the control file name (i.e., if the name of the control file is `HIVenvSweden_0.ctl`, type `codeml  HIVenvSweden_0.ctl`) to run the program. Model 0 runs pretty quickly, but Model 1 can take anywhere from a few minutes to a few days, depending on the number of taxa and the speed of your computer. By opening two Terminal/Command Prompt windows you can run the `HIVenvSweden_0.ctl` and `HIVenvSweden_1.ctl` files simultaneously, but you will need to put the .ctl files into separate folders and put copies of the sequence and tree files into both folders. If you run two .ctl files sequentially, be sure to move the output files from the first run to a different folder before doing the second run lest they be overwritten.

Identify the branches along which selection may have occurred

At the end of the run several files are written, but we are only interested in the `.out` files, `HIVenvSweden _0.out` and `HIVenvSweden _1.out`. Use any text editor to open `HIVenvSweden _1.out` and scroll to the bottom of the file to see **Figure 14.12**.

The table at the bottom of the `out` file gives the dN/dS ratio for each branch and also the number of silent changes (`S*dS`) and replacement changes (`N*dN`). The branches are given by the numbers of the nodes at each end of the branch.

The terminal nodes, corresponding to the sequence names, are numbered in the order that they appear at the top of Figure 14.12. Thus U68496 is 1, U68497 is 2, and so forth. To number the nodes, work backward from the terminal taxa. Node 17, for instance, is the node that connects to both U68496 and U68497. Observing that node 13 connects to three nodes (nodes 6, 14 and 18) in Figure 14.12 may be surprising, but remember that the ML tree is unrooted, so what appears to be the root doesn't really exist.

```
U68496
U68497                0.5526 (0.0287 0.0519)
U68498                3.6604 (0.0925 0.0253) 4.5034 (0.1140 0.0253)
U68499                3.7961 (0.0974 0.0257) 4.2041 (0.1081 0.0257)-1.0000 (
U68500                2.9760 (0.1298 0.0436) 3.2244 (0.1409 0.0437) 1.8847 (
U68501                2.7890 (0.1194 0.0428) 3.3029 (0.1417 0.0429) 5.7001 (
U68502                1.5754 (0.1631 0.1035) 1.8856 (0.1777 0.0942) 1.8749 (
U68503                2.4325 (0.1475 0.0606) 2.4262 (0.1474 0.0607) 2.9485 (
U68504                1.9018 (0.1331 0.0700) 1.3384 (0.1443 0.1078) 1.1433 (
U68505                1.4342 (0.1274 0.0888) 1.5561 (0.1385 0.0890) 1.3640 (
U68506                3.1697 (0.1360 0.0429) 1.5016 (0.1473 0.0981) 1.5238 (
U68507                1.9603 (0.1535 0.0783) 2.2539 (0.1769 0.0785) 1.6424 (
```

Figure 14.12

```
TREE #  1:  ((((4, 5), 3), (1, 2)), 6, (7, (8, (12, (10, (9, 11))))));  MP
lnL(ntime: 21  np: 43):  -1029.664282      +0.000000
   13..14    14..15    15..16    16..4    16..5    15..3    14..17    17..1     1
0.053086 0.023449 0.088060 0.024903 0.102862 0.043714 0.150625 0.029294 0.

Note: Branch length is defined as number of nucleotide substitutions per co

tree length =    1.58754

((((4: 0.024903, 5: 0.102862): 0.088060, 3: 0.043714): 0.023449, (1: 0.0292

((((U68499: 0.024903, U68500: 0.102862): 0.088060, U68498: 0.043714): 0.023

Detailed output identifying parameters

kappa (ts/tv) =   2.96179

w (dN/dS) for branches:  0.63120 999.00000 999.00000 999.00000 1.02794 999.

dN & dS for each branch

branch          t        N        S   dN/dS     dN      dS    N*dN   S*dS

   13..14    0.053    231.0    42.0  0.6312  0.0162  0.0257    3.8    1.1
   14..15    0.023    231.0    42.0 999.0000 0.0092  0.0000    2.1    0.0
   15..16    0.088    231.0    42.0 999.0000 0.0347  0.0000    8.0    0.0
   16..4     0.025    231.0    42.0 999.0000 0.0098  0.0000    2.3    0.0
   16..5     0.103    231.0    42.0  1.0279  0.0344  0.0335    8.0    1.4
   15..3     0.044    231.0    42.0 999.0000 0.0172  0.0000    4.0    0.0
   14..17    0.151    231.0    42.0 999.0000 0.0593  0.0001   13.7    0.0
   17..1     0.029    231.0    42.0  0.0965  0.0040  0.0415    0.9    1.7
   17..2     0.076    231.0    42.0  0.5375  0.0224  0.0416    5.2    1.7
   13..6     0.129    231.0    42.0 999.0000 0.0506  0.0001   11.7    0.0
   13..18    0.034    231.0    42.0 999.0000 0.0135  0.0000    3.1    0.0
   18..7     0.193    231.0    42.0  1.2786  0.0665  0.0520   15.4    2.2
   18..19    0.056    231.0    42.0  0.3057  0.0137  0.0449    3.2    1.9
   19..8     0.228    231.0    42.0  1.3522  0.0793  0.0586   18.3    2.5
   19..20    0.066    231.0    42.0  0.3896  0.0176  0.0452    4.1    1.9
   20..12    0.065    231.0    42.0  0.7227  0.0205  0.0284    4.7    1.2
```

With the nodes numbered (**Figure 14.13A**), we can write the dN/dS ratio above each of the branches (**Figure 14.13B**). Notice that seven branches in Figure 14.13B don't have dN/dS ratios indicated. That is not an omission. In Figure 14.12, each of those branches is indicated as having a dN/dS ratio of 999.0000, which is PAML's way of indicating a ratio of infinity. In each case there are zero silent substitutions (S*dS column in Figure 14.12) along that branch. Similarly, the branch from node 22 to U68504 is shown as having a dN/dS ratio of 0.0001—PAML's way of indicating zero—because the number of replacement substitutions (N*dN) is zero.

Figure 14.13

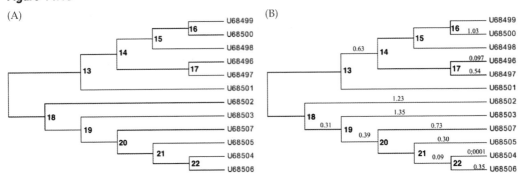

There are several branches that were under purifying selection (dN/dS <1) and several that were under positive selection (dN/dS >1). There appears to be a lot of variation in both the intensity and direction of selection.

Test the statistical significance of the dN/dS ratios

We need to know how much we can trust the conclusion that selection was variable, and how much we can trust the conclusions of positive selection on those branches where dN/dS appears to be >1 and purifying selection on those branches where dN/dS was <1.

Use the likelihood ratio test to test the significance of estimating individual dN/dS ratios versus estimating one ratio for the entire tree. The red arrow in Figure 14.12 indicates the line where the log likelihood is shown. The log likelihood for Model 1 is –1029.664282. If we open the Model 0 output file and look at the same line, we see that the log likelihood is –1044.954861 for Model 0 (where dN/dS was estimated as 0.8331 for all branches). Since –1029.664282 is greater than –1044.954861, Model 1 is indeed more likely than Model 0.

A **likelihood ratio test** will tell us if the Model 1 likelihood is *significantly* greater than the Model 0 likelihood. We first calculate *twice* the difference between the likelihoods as 30.581158 and use that as the chi-square value in a chi-square test. The degrees of freedom is the difference in the number of parameters estimated under the two models. Model 0 estimates one parameter (the mean dN/dS ratio) and Model 1 estimates one parameter for each of the 21 branches, so there are 20 degrees of freedom. (A handy chi-square calculator can be found at http://www.fourmilab.ch/rpkp/experiments/analysis/chiCalc.html). For 20 degrees of freedom and a chi-square of 30.581158, the p value is 0.0609. Because $p > 0.05$, we can conclude that variable dN/dS ratios are almost—but not quite—significantly better than one homogeneous dN/dS ratio.

This still does not tell us about the reliability of any particular branch's ratio. When the number of changes is small, it is possible for there to be an excess of replacement substitutions by chance alone. We can compare two versions of Model 2 to ask whether the dN/dS ratio of the branch from node 19 to U68503 (see Figure 14.13B) is significantly greater than 1. Model 2 estimates one ratio for a branch of interest, and another homogeneous ratio for the rest of the branches. We mark a branch by inserting **#1** immediately before the colon that sets off that branch's length in the tree file. Look at **HIVenvSweden_19.nwk** to see the **#1** that marks the branch leading to **U68503** (i.e., the branch with the dN/dS ratio of 1.35).

Download Chapter 14: HIVenvSwedenML_19.nwk

Let's designate the two versions of Model 2 (see p. 214) as Model 2A and Model 2B. We will modify the .ctl files in a manner similar to setting up the Model 0 and Model 1 runs.

Download Chapter 14: HIVenvSweden_2A.ctl

Download Chapter 14: HIVenvSweden_2B.ctl

For Model 2A, I set **model = 2** and **fix_omega = 0** in the control file. The latter setting means that the program will estimate one dN/dS ratio for the marked branch and a different homogeneous or background ratio for all the other branches. For Model 2B, I set **model = 2** and **fix_omega = 1**. This setting means that the program will set a dN/dS ratio of 1.0 for the marked branch and estimate the background ratio for the other branches. The outfile for Model 2A estimates dN/dS for that branch as 1.3041; the log likelihood for the model is –1044.746055. The log likelihood for Model 2B is –1044.881314. Model 2A estimates two parameters (one ratio for the marked branch and another for all other branches) and Model 2B estimates one parameter (the background ratio), so for the likelihood ratio test chi-square is 0.27 for one degree of freedom, and $p = 0.6033$. The dN/dS ratio of that branch is not significantly greater than 1. If we instead mark the branch from node #18 to U68502, we could run Models 2A and 2B again to see whether that branch is under significant positive selection.

Summary

Pairwise comparisons of HIVenvSweden sequences (see Figure 14.4) showed several pairs for which positive selection appeared to have taken place as they diverged from each other. On the other hand, there was no evidence for positive selection operating on any of the individual codons (see Figure 14.10). In addition, evidence for variable dN/dS ratios along the branches was not quite significant, and even those branches that showed the largest positive dN/dS ratios were not significantly different from 1.0. Overall, then we are not justified in concluding that the *env* gene of HIV was evolving under positive selection in the population from which this sample came.

Phylogenetic Networks

Why Trees Are Not Always Sufficient

In the early days of molecular evolution, before DNA sequencing was possible, protein sequences were used to estimate the evolutionary histories of different species. The sequencing targets were small, ubiquitous proteins such as cytochrome *c* (small because protein sequencing was difficult and expensive). The resulting phylogeny was seen as "the" phylogeny of that set of species. As time went on, and especially as DNA sequencing became commonplace, investigators began to sequence numerous different proteins and genes from the same set of species. Not surprisingly, the resulting phylogenies didn't always agree with one another.

Sometimes the disagreement could be attributed to uncertainty in various parts of the trees, but eventually it became obvious that there were cases of equally strongly supported phylogenies that differed depending on which gene was sequenced. This led to cautions about the difference between a gene tree and a species tree, but with little sense of what was meant by a "species tree." Was a species tree, after all, just a well-accepted tree based on morphology? Was it the tree supported by the majority of the genes? Or was it some sort of average among the different trees?

It became necessary to accept that different robust trees from different genes of the same set of species meant that *the different genes had different evolutionary histories*. Now that horizontal gene transfer and incomplete genetic isolation early in the speciation process are well-accepted concepts, the notion of different evolutionary histories for different genes is no longer controversial, or even surprising. Indeed, in this era of rapid and inexpensive genome sequencing, with thousands of genes from the same set of species to compare, the evidence for differing evolutionary histories for individual genes is so overwhelming that it hardly merits much discussion.

Still, we are faced with the problem of trying to describe a process that applies to a set of species as a whole in the face of conflicting, but equally reliable, evidence. In a sense, a phylogenetic tree is a naïve attempt at a simple explanation (mutation and genetic isolation) for a complex process (speciation). A phylogenetic tree assumes complete genetic isolation as species evolve by a process of mutation, but the conflicts among trees is direct evidence for incomplete genetic isolation. Indeed, there are a variety of factors—including gene duplication and loss, hybridization, recombination, and horizontal gene transfer—that require a more complex model of evolution than is provided by a phylogenetic tree. When a single phylogenetic tree is insufficient to describe the evolutionary history of a set of species, we require an alternative. That alternative is a phylogenetic network. Like a tree, a phylogenetic network is a graph that describes an evolutionary process. A network, however, diagrams the conflicting information as well as the consistent information and thus illustrates the alternative histories for different parts of a data set.

While I have cast this discussion in terms of conflicts between the histories of different genes, it applies equally well to conflicts between the histories of different parts of the same gene. Any process, be it horizontal transfer, hybridization, gene duplication and loss, or recombination, that can lead to different histories for different parts of a genome can lead also to different histories for different parts of a gene. Indeed, such conflicts sometimes result in uncertainty even within a tree of a single gene.

An important role of phylogenetic networks is in understanding evolution *within* a species. For sexual species there is no genetic isolation within the species except for such special cases as mitochondrial DNA (inherited strictly from the maternal parent) and, in some cases (humans, for instance), genes on the Y-chromosome (strictly male, no recombination). Even supposedly asexual organisms such as bacteria exhibit evidence of recombination that can make phylogenetic trees invalid. Phylogeographic studies of a species are much better suited to phylogenetic networks than phylogenetic trees.

This chapter is a necessarily superficial discussion of phylogenetic networks (see *Learn More about Phylogenetic Networks*, p. 223). The material presented here is drawn almost entirely from the new book *Phylogenetic Networks: Concepts, Algorithms and Applications* by D. H. Huson, R. Rupp, and C. Scornavacca (2011). If this chapter suggests to you that phylogenetic networks will be useful in your work you are strongly advised to read that book.

Unrooted and Rooted Phylogenetic Networks

Just as there are rooted and unrooted phylogenetic trees, there are rooted and unrooted phylogenetic networks. It turns out that it is much more difficult to estimate rooted phylogenetic networks directly from sequence data than it is to estimate unrooted phylogenetic networks; indeed methods for estimating rooted networks directly are not yet well developed. That should be no more disturbing than recognizing that all of the phylogenetic methods discussed in this book estimate unrooted trees. Although we are usually more interested in rooted than in unrooted trees, we root those trees with an outgroup. Similarly, we can root unrooted phylogenetic networks with an outgroup. SplitsTree is a program for estimating phylogenetic networks from aligned DNA and protein sequences.

There are good methods for combining conflicting rooted phylogenetic trees to make rooted networks that show the conflicts among those trees. Fortunately, there is also a software package, Dendroscope, for estimating rooted phylogenetic networks from sets of rooted phylogenetic trees. As a practical matter we will therefore consider estimation of unrooted and rooted phylogenetic networks as separate issues.

Using SplitsTree to Estimate Unrooted Phylogenetic Networks

Estimating networks from alignments

SplitsTree includes an excellent manual that certainly deserves your attention. I will not attempt to recapitulate that manual here. SplitsTree can read alignments in FASTA, PHYLIP, or Nexus formats. The MiniData alignment—the SmallData data set with all the Gram positive strains except *G. vaginalis* removed—serves here to illustrate the use of SplitsTree.

 Chapter 15: MiniData.nxs

Figure 15.1

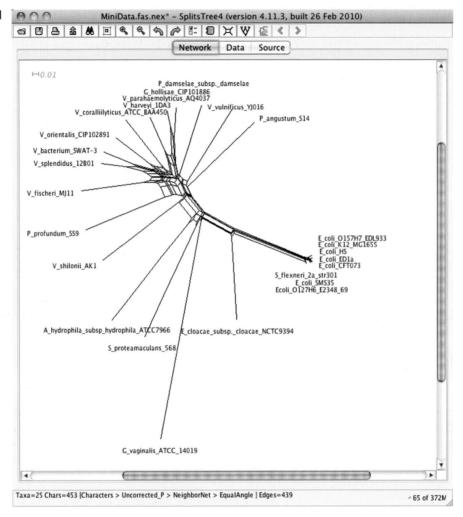

When an alignment file is opened, SplitsTree automatically estimates and displays a network using the NeighborNet algorithm (**Figure 15.1**). The status line at the bottom of the window tells us that the network was estimated from an alignment of 453 characters for each of 25 taxa; that the distances were calculated by the uncorrected (observed) proportion of changes, that the network was estimated by the NeighborNet method, and that the network has 439 edges.* The information on the status line can be set in the **Edit>Preferences** menu choice.

---

*Mathematicians think of graphs that have nodes and edges, while biologists think of trees that have nodes and branches. Because most of the work on phylogenetic networks is done by mathematicians, documentation of network software typically

Phylogenetic Networks

Unrooted Networks

It is easiest to begin this description of phylo-genetic networks by thinking about something familiar: an unrooted phylogenetic tree. An unrooted tree is just a particular case of an unrooted network. Such a tree is an undirected graph in which each interior node has at least three branches connected to it. The branches cannot be said to lead toward or away from a node because, being unrooted, we know nothing about directionality.

A *split* is the result of removing a branch, and a split divides the set of taxa into two mutually exclusive sets. A *trivial split* is one that removes a branch that connects to a single external (leaf) node; here we are concerned with *non-trivial splits*, which remove a branch that connects two interior nodes. Removing the branch represented by the dashed line in Figure 1A splits the tree into two sets of taxa {AB} and {CDE}. This non-trivial split S can also be represented as AB|CDE or as $\frac{AB}{CDE}$ (or CDE|AB or $\frac{CDE}{AB}$).

If the complete set of splits—that is, every split that can be obtained by removing exactly one branch—is *compatible*, then that set of splits can be represented by a phylogenetic tree. In addition to the split $S_1 = AB \mid CDE$, there is a second non-trivial split in Figure 1A: $S_2 = ABC \mid DE$. A split S_i can be generalized as $S_i = X_i \mid Y_i$, with X_i and Y_i being subsets of S_i. A pair of splits S_1 and S_2 are compatible if *at least one* intersection of the subsets is empty. For the case shown in Figure 1A, the intersection

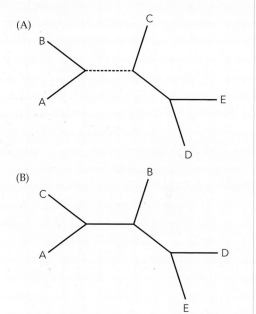

FIGURE 1 Two incompatible unrooted trees.

of X_1 (AB) and X_2 (ABC) is any taxa that appear in both subsets, in this case {A,B}. Similarly the intersection of X_2 and Y_1 is C, and that of Y_1 and Y_2 is D. The intersection of X_1 (AB) and Y_2 (DE), however, is empty, and splits S_1 and S_2 are therefore compatible. Since there are only two non-trivial splits, all of the splits are compatible and Figure 1A is a tree.

Similarly, an unrooted network is an undi-rected graph in which incompatible splits are

---

uses the term edges. For a tree the terms *edge* and *branch* are synonymous, but for networks *edges* seems more appropriate and I will usually use that term when discussing networks. Like branches, edges are represented by lines and connect nodes.

represented by parallel edges. (As noted in the footnote on p. 223, I use the term *edges* when referring to networks and *branches* when referring to trees, but in practice the two terms are synonymous.) The parallel edges represent alternative ways to connect internal nodes—in other words, alternative evolutionary trajectories. Suppose that some data support the tree in Figure 1A, but other data support the tree in Figure 1B. That inconsistency can be shown as the network in Figure 2. Splits are obtained by removing all of the parallel edges. Figure 2 has two sets of parallel edges, one shown in red the other in blue. Removal of the blue edges gives S_1 = AC | BDE, and removal of the red edges gives S_2 = AB | CDE. The intersections X_1Y_1 = A, X_1Y_2 = C, X_2Y_1 = B, and Y_1Y_2 = DE. There is no empty intersection, so S_1 and S_2 are incompatible and Figure 2 does not show a tree but instead shows a network.

A fully resolved unrooted tree of N taxa has $N - 3$ internal nodes, but an unrooted network can have almost any number of nodes and edges. The more complex a network is, the more incompatibilities it reveals—but at the same time the more difficult it becomes to take in and interpret those inconsistencies. (I use the term *complex* in the visual, not the mathematical, sense.) Methods that seek to reduce the complexity of a network do so at the expense of detail, but they often make the resulting graphic more useful.

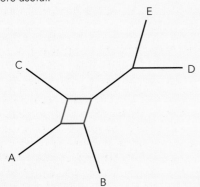

FIGURE 2 An unrooted network incorporating the inconsistencies between the trees shown in 1A and 1B.

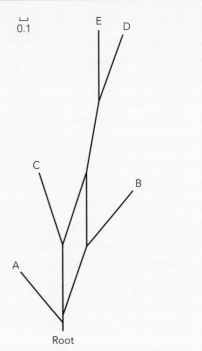

FIGURE 3 Network from Figure 2 rooted on taxon A.

Just as it is possible to root an unrooted tree using an outgroup, it is possible to root an unrooted network. Figure 3 shows the network of Figure 2 rooted on taxon A using the program SplitsTree.

Phylogenetic Networks from Sequence Alignments

As a practical matter, the issue is not how to determine splits from a tree or a network; it is how to estimate splits from the data and then create a tree or network that is consistent with those splits. There are a variety of methods available to achieve these goals, only a few of which will be discussed here.

DISTANCE METHODS Just as the NJ method for trees first constructs a distance matrix then uses that distance matrix to make a tree, distance methods for networks construct a distance matrix and use it to estimate splits.

Split-decomposition produces a set of weighted splits that are weakly compatible, a method that ensures that the resulting network will not be too complicated. It is a very conservative method that tends to exhibit low resolution and should be limited to data sets of fewer than 100 taxa.

Neighbor-net produces a set of weighted splits that is circular. The neighbor-net method is more popular than split-decomposition because it is less conservative and does not tend to lose resolution on larger data sets. The resulting network can be drawn without edges crossing.

SEQUENCE METHODS   Sequence methods estimate splits directly from a sequence alignment—or, more accurately, from a condensed sequence alignment. A condensed alignment eliminates identical sequences, eliminates all constant columns, and none of the remaining columns exhibit the same pattern of states (although the number of columns with the same pattern is kept track of as a means of weighting the splits).

The **parsimony splits method** aims to find at most two most-parsimonious topologies for any given set of four taxa and calculates an index of each resulting split. The set of indices are processed by an algorithm such as the convex hull algorithm to make a network. Because the result is usually very similar to that of split-decomposition, the parsimony splits method is not widely used.

Median networks were developed to deal with closely related sequences and are best applied to closely related sequences that have evolved without recombination (such as Y chromosomes from human populations) or to phylogeographic studies. Median networks are estimated from binary data (i.e., data in which characters can have only two possible states). Because sequence data obviously has more than two character states, one way to convert sequence alignments to binary data is to further condense the alignment by eliminating all columns in which there are more than two states. From each possible set of three sequences in

that condensed alignment, a median sequence is calculated by using the most frequent character state at each site in the triplet of sequences. The median network is calculated from that set of median sequences. This approach means discarding a lot of the information. In answer to this objection, however, note that if the sequences are indeed closely related, most columns will either be constant (and thus eliminated) or will be binary.

Because median networks can have a large number of nodes and edges even for a small number of taxa (especially if there are many parallel or back-mutations), *reduced median networks* attempt to simplify matters by identifying likely parallel mutations and making them explicit.

A *quasi-median network* deals directly with multistate data instead of eliminating multistate columns. Whenever a column of three different states occurs in the computation of the median, three different median sequences are produced. Still, the number of nodes and edges can be very large, and a *pruned quasi-median network* is an attempt to further reduce complexity so that the resulting network consists only of those pairs of nodes and edges that differ by exactly one position.

Median joining attempts to estimate a network that is as informative as a quasi-median network but is even simpler. Median joining is a heuristic method based on relaxed minimum-spanning networks, a subject that is beyond this brief discussion.

Phylogenetic Networks from Trees

You are already familiar with the concept of consensus trees, such as a consensus of a set of bootstrap trees (NJ and ML) or the consensus of a set of post-burnin Bayesian trees. Consensus trees are determined from the splits associated with each tree. The strict consensus is based on the set of splits that occur in every tree, while the majority rule consensus is based on the set of splits that occur in >50% of the trees. A consensus tree displays those parts of the evolutionary history upon which the trees (or >50% of the trees) agree; those parts of the history that are incompatible are suppressed. In contrast, a phylogenetic network attempts to represent the inconsistent as well as the consistent parts of the evolutionary history.

The program SplitsTree implements all three of the following network types.

CONSENSUS NETWORKS   If Sp is the set of splits that are present in proportion p of the trees and $p \geq 0.5$, the splits will be compatible and the result will be a consensus tree. If $p < 0.5$, some of the splits may be incompatible and the result will be a phylogenetic network. In general, the visual complexity (number of parallel edges) of a consensus network will increase as p decreases. For even modest numbers of taxa it is worth trying several values of p to determine which consensus network is most useful under the circumstances. For example, Figure 2 was obtained by making a consensus network from the trees in Figure 1 using a threshold of 0.33.

SUPERNETWORKS   The calculation of consensus networks requires that all the trees contain identical taxa. But what if the trees don't have identical sets of taxa? You may have trees for 20 different species based on 10 different genes, but have data for only 12–15 species for each gene. A supernetwork allows you to combine that set of trees in order to get an overall view of the evolution of that set of taxa. In effect, the method uses the number of trees in which a split occurs as a filter to determine which splits make it into the network. For studies within a species, supernetworks allow you accommodate sets of genes in which some genes may not be present in all individuals (e.g., bacterial genomes, where typically only a minority of the genes in a given species are present in all individuals).

FILTERED SUPERNETWORKS   An alternative to using the number of trees in which a split occurs as the filter is to assign a distortion value to each split. The distortion value measures how much each input tree would have to be modified to accommodate the split, and provides a parameter for controlling the complexity of the supernetwork.

Rooted Networks

In general biologists are more interested in rooted networks than in unrooted networks because rooted networks tell us about the direction of evolution. A rooted phylogenetic tree (see Figure 4) is a special case of a rooted network, and is a directed graph in which all but one node has one branch leading toward the node, two or more branches leading away from interior nodes, and zero branches leading away from exterior (leaf) nodes. Exactly one node, designated, the root node, has zero branches leading to it. As a result all paths lead inexorably from interior nodes toward leaf nodes. Biologically those branches represent mutations that have occurred during evolution.

Removal of an internal branch results in a subtree whose root is the node toward which

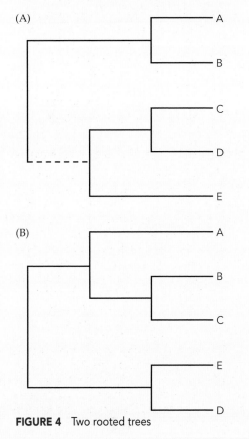

FIGURE 4   Two rooted trees

the removed branch pointed. The taxa that are included in that subtree constitute a cluster or a clade, all of which are descended from the ancestor represented by the root of the subtree. Removal of the dashed branch in Figure 4A produces a subtree whose taxa are the clade (cluster) {CDE}.

If we have a set of rooted phylogenetic trees, the purpose of a cluster network would be to illustrate the parts of the phylogeny that agree and the parts that are resolved in different ways. The methods in this section involve calculating rooted networks from the clusters in sets of rooted trees using the program Dendroscope.

The methods are based on the concept of compatible and incompatible clusters, just as unrooted networks are based on the concepts of compatible and incompatible splits. A pair of clusters are compatible if (1) one is a subcluster of the other, or (2) there is no overlap in membership between the two clusters. Thus in Figure 4A the clusters {CD} and {CDE} are compatible because {CD} is a subcluster of {CDE}. Likewise, {CDE} and {AB} are compatible because there is no overlap in membership. However, cluster {CDE} of Figure 3A and {ABC} of Figure 4B are not compatible because they both contain C.

Hardwired versus Softwired Networks

Hardwired networks contain every cluster that is present in any of the trees that make up the network. The trees in Figure 4 contain the following clusters: ABCDE | AB | CDE | CD | ABC | DE | A | B | C | D | E.

In a hardwired network, each edge defines a cluster. A reticulation occurs at a node if there are more than one edge coming into that node. The hardwired network in Figure 5A has two reticulate nodes, C and D. Every node that is present in the two trees shown in Figure 4 is present in the hardwired network in Figure 5A.

Consensus clustering is an example of the methods used to produce hardwired networks. If we have a set of rooted phylogenetic trees, a cluster network can be computed by a cluster-popping algorithm in which all of the clusters in the input set of trees are defined, then assem-

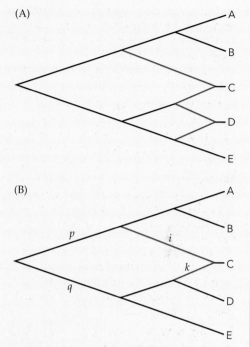

FIGURE 5 (A) Hardwired and (B) softwired networks of the two trees shown in Figure 4.

bled into a network according to a set of rules. This works well when the number of incompatibilities among the trees is small, but in more complicated cases it is necessary to include only those clusters that occur in a certain percentage of the trees, thus obtaining a consensus cluster network. As was the case for unrooted consensus networks, this method has the disadvantage of showing only those differences that are supported by that percentage of the trees; clusters that occur on fewer trees are not represented.

In a *softwired* network, an edge may define more than one cluster and different methods must be applied. To obtain the clusters from a softwired network, the edges leading into a reticulate node are treated as alternatives in which one edge is "on" and the others are "off." In the softwired network seen in Figure 5B there is one reticulate node, C. If edge *i* is on and edge *k* is off, then edge *p* defines the cluster ABC, edge *q* defines the cluster DE, and the network becomes the tree in Figure 4B. If edge *i* is off and edge *k* is on, edge *p* defines the cluster AB, edge *q* defines the cluster CDE, and the network in Figure 5B becomes the tree in Figure 4A. Edges *p* and *q* thus each define two clusters.

Because edges of softwired networks can define more than one cluster, softwired networks usually contain fewer nodes and edges than do hardwired networks of the same trees and are thus less visually complex. In this example turning on and off the edges leading to the reticulate nodes recovers each of the trees that make up the network, but that is not always the case. It is the case, however, that turning on and off the various possible combinations of edges leading to reticulate nodes always recovers all of the clusters that make up the input trees.

Galled networks are softwired networks that seek the minimum number of reticulations. Because the method is heuristic it does not guarantee to find the best solution. **Minimal level-k** networks are softwired networks that seek the minimum number of levels in the network. The number of levels is the maximum number of reticulations in any biconnected component of the tree. The minimal level method is also heuristic and does not guarantee the optimal solution. ◀▮

The interpretation of the unrooted phylogenetic network is certainly not intuitive. It is helpful to read pp. 223–224 of *Learn More about Phylogenetic Networks* to understand what is being shown in Figure 15.1. The parallel lines are edges that indicate alternative inconsistent evolutionary trajectories. In an unrooted tree, removal of an internal branch splits the taxa into two mutually exclusive sets. Similarly, removal of a set of parallel edges in a network divides the taxa into two sets and is the equivalent of a split for unrooted phylogenetic networks. Clicking any edge selects that edge and all of its parallel edges plus all of the nodes in the smaller set (red in **Figure 15.2**).

⊢0.01

Figure 15.2

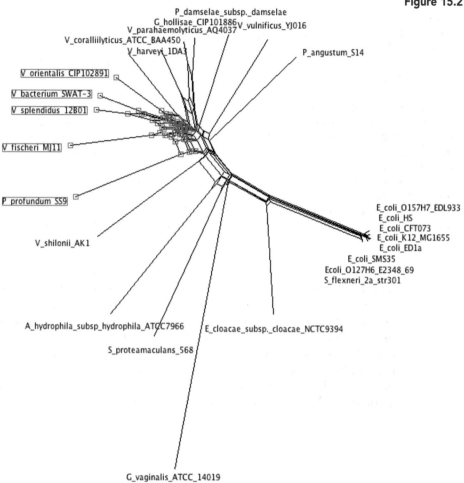

Use the mouse scroll wheel or the magnifying glass icon in the toolbar at the top of the window (see Figure 15.1) to zoom in. The parallel edges are shown in red, all of the nodes that are split off by removal of those edges are indicated by small red squares, and all of the taxa in that subset are shown in red boxes (**Figure 15.3**).

Figure 15.3

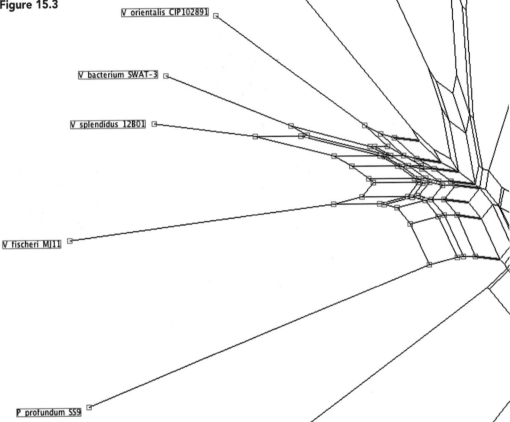

All of those parallel edges give a visual picture of the conflicting information that is masked by the simpler appearance of the corresponding phylogenetic tree. Compare Figures 15.2 and 15.3 with **Figure 15.4** and notice, for example, how Figure 15.4 (the corresponding phylogenetic tree) conceals the conflicts among *P. profundum,* *V. fischeri,* and *V. splendidus.*

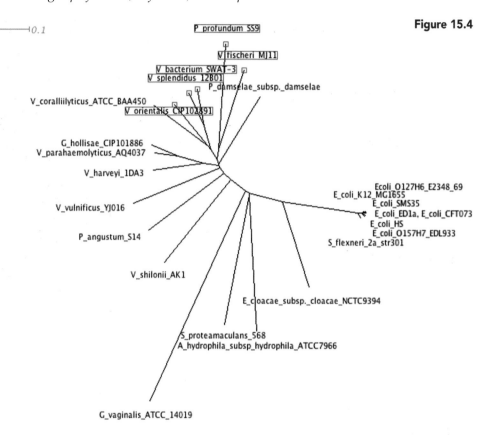

Figure 15.4

Figure 15.5

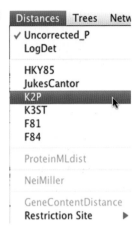

SplitsTree provides a variety of substitution models to calculate the distances upon which the network is based. Choose an alternative model from the **Distances** menu (**Figure 15.5**).

Figure 15.6 shows the network when the K2P model is used to calculate distances. That network is simpler and has only 418 edges compared with the Uncorrected P-based network seen in Figure 15.1. More complex models such as HKY produce a star phylogeny rather than a network.

Figure 15.6

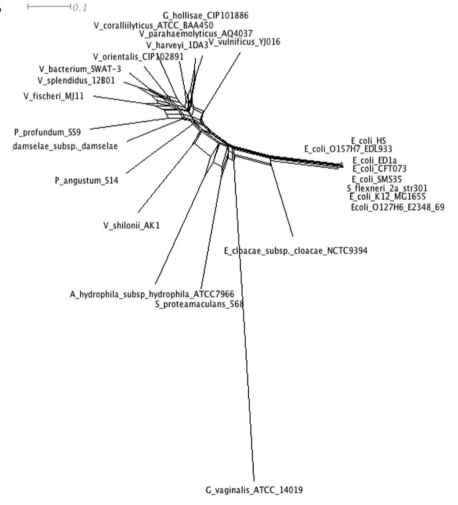

The **Networks** menu (**Figure 15.7**) offers a variety of alternative methods for estimating the network, some of which (e.g., median networks) are not suitable for this data set.

It is just as important to estimate reliability of a network as that of a tree. Choose **Bootstrap** from the **Analysis** menu to estimate the bootstrap support for the splits. Clicking an edge selects all of the parallel edges and the support for the split that is induced by removal of those edges (**Figure 15.8**).

Figure 15.7

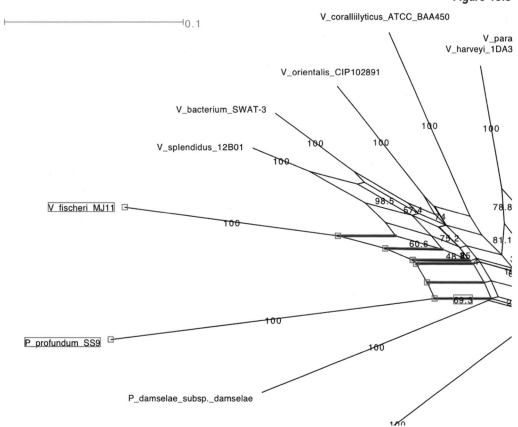

Figure 15.8

Rooting an unrooted network

Choose **Rooted Equal Angle** from the **Draw** menu to root the network by an outgroup. The **Processing Pipeline** window allows you to specify one taxon as the outgroup. **Figure 15.9** specifies the Gram positive *G. vaginalis* as the outgroup. Clicking the **Apply** button produces the rooted network depicted in **Figure 15.10**.

Figure 15.9

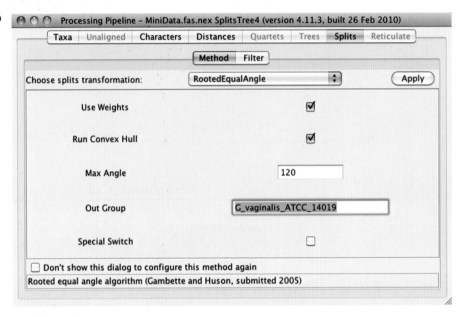

The format shown in Figure 15.10 gives direction to the evolutionary picture and shows that the *E. coli/Shigella/E. cloacae* group diverged from the *Vibrio* and other Gram negative taxa early on, but has undergone less divergence within that clade than has the *Vibrio* et al. clade. At the same time, the conflicting information is there for all to see. Indeed, this network image makes it clear that speciation is not an instantaneous process that takes place at a single point in time, but is a complex process occurring over a period of incomplete genetic isolation.

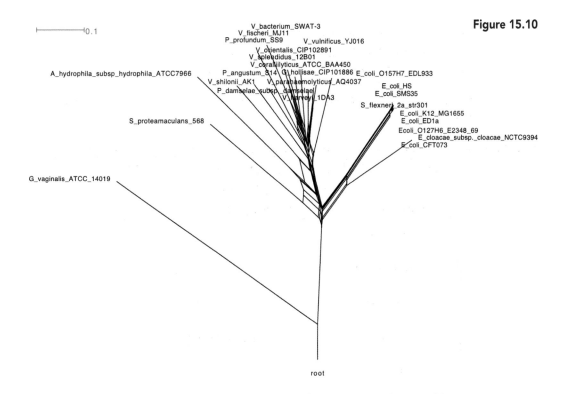

Figure 15.10

Estimating networks from trees

The input is a file of trees in either the Newick or the Nexus format. A Nexus file of the post-burnin trees from a BI estimate based on the 27 "reliable sequences" of the SmallData data set (see Figure 12.13B) included 5622 trees. Keep in mind that those trees all contain exactly the same taxa because they are roughly equally likely trees estimated by the Bayesian Inference method from one data set.

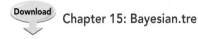 Chapter 15: Bayesian.tre

Consensus networks

When the file is opened, SplitsTree automatically calculates a **consensus network** based on a threshold of 0.3—that is, a split must occur in 30% of the trees in order to be included in the network. For the current data set, that network has 51 edges (**Figure 15.11**).

Figure 15.11

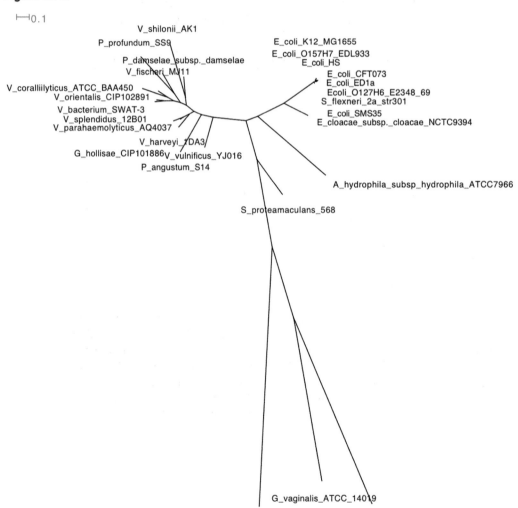

A fully resolved unrooted tree of N taxa has $2N-3$ branches (edges), meaning that the Figure 15.11 is indeed a tree and that there are no conflicts among the splits that occur in more than 30% of the trees. This consensus tree, however, ignores those splits that occur in fewer trees.

To lower that threshold, choose **ConsensusNetwork** from the **Networks** menu to reveal the **Processing Pipeline** window (**Figure 15.12**).

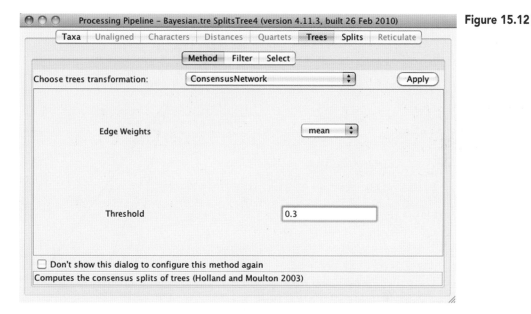

Figure 15.12

Figure 15.13 shows the results of progressively lowered thresholds. Setting the threshold to 0.1 results in the network in Figure 15.13A; setting it to 0.03 produces the network in Figure 15.13B, and a threshold of 0.01 leads to the network in Figure 15.13C. As the threshold is lowered, more incompatible splits are included in the network and the network becomes more complex. Setting the threshold to 0.0 reveals all the incompatible splits but produces a network so complex it is almost impossible to interpret (Figure 15.13D).

Figure 15.13

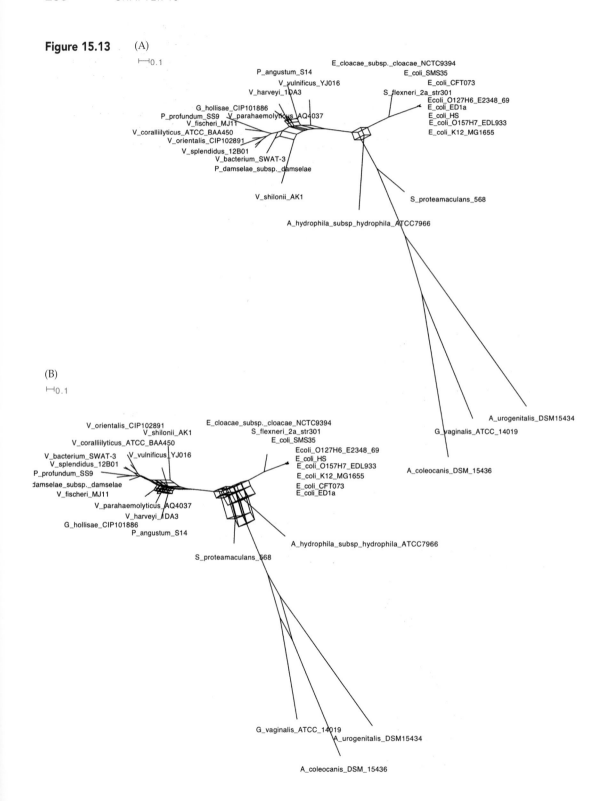

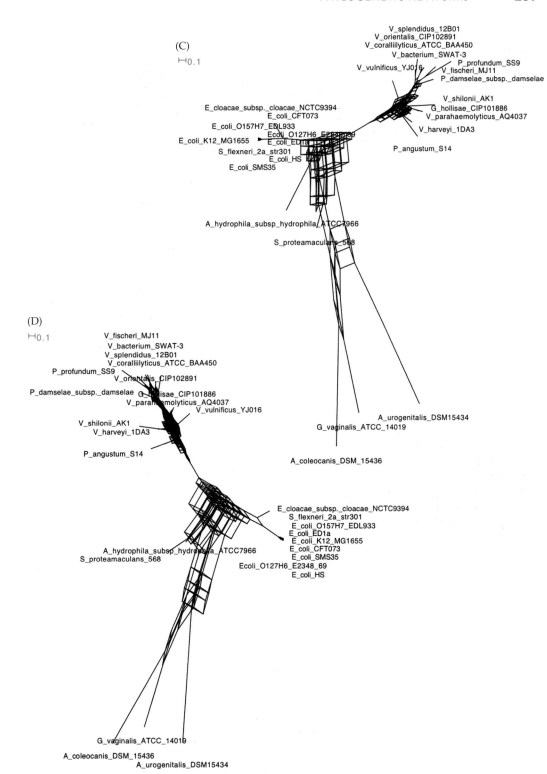

There are several choices for setting the edge weights in a consensus network (**Figure 15.14**). The default is **mean**, in which edges are weighted according to the mean lengths of the branches of the trees. Choosing **none** produces the equivalent of a cladogram (i.e., all edges are the same length). Choosing **count** or **sum** produces a network in which the edges are drawn according to the frequency with which they occur in the set of trees.

Figure 15.14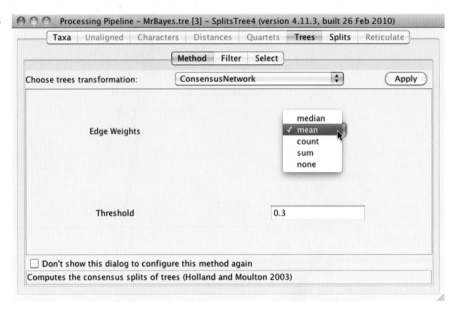

Supernetworks

Consensus networks require that all trees include exactly the same set of taxa. When that condition does not apply, choose **SuperNetworks** from the Networks menu. The supernetwork method, however, is not just a solution to the problem of different taxa sets (that method, which uses the *Z-closure algorithm*, produces a network that is different from the consensus network even when all the trees include the same taxa). **Figure 15.15** shows the supernetwork of the same trees used to make the consensus network in Figure 15.11 (i.e., with a threshold of 0.3).

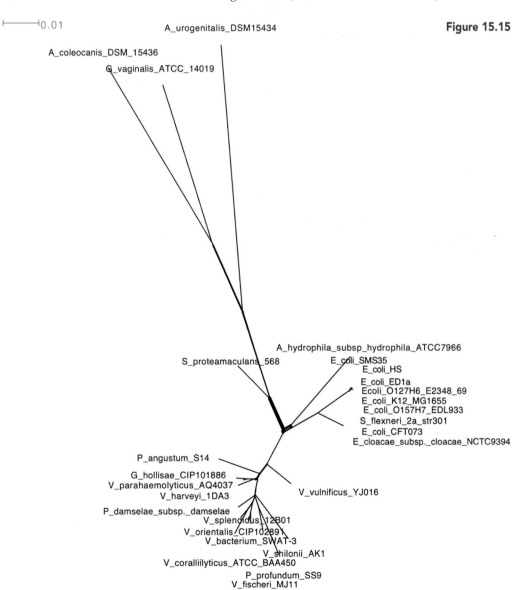

Figure 15.15

At first glance the supernetwork seems to show the same treelike structure as Figure 15.11. However, the supernetwork includes 1262 edges compared with 51 for Figure 15.11. A close-up view of the center of that network shows the multiple edges (**Figure 15.16**).

Figure 15.16

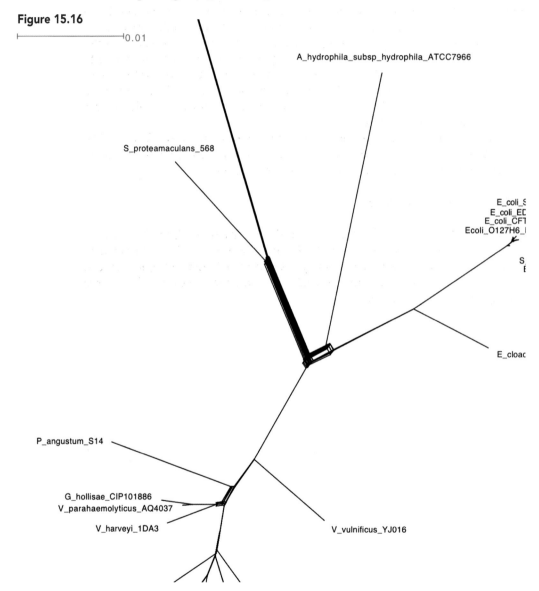

Using Dendroscope to Estimate Rooted Networks from Rooted Trees

Chapters 6, 8, 9, and 10 discussed estimating phylogenetic trees by four different methods, and the resulting trees differed slightly from one another. I pointed out that these differences can be thought of as reflecting real uncertainty and suggested that, as long as the trees are robust, it doesn't matter much which tree you choose to present to your audience. One alternative to choosing is to present all the trees, but perhaps a better alternative is to present those trees, including their conflicts, in the form of a network. This alternative has the advantage of making it clear where there are conflicts and where there is agreement among the trees.

Dendroscope can read tree files in both the Newick and Nexus formats. The file `All.tre` includes the NJ, Parsimony, ML, and BI trees estimated from the reliable columns of the SmallData data set (see Chapter 12).

 Chapter 15: All.tre

Figure 15.17 shows the NJ tree (A), the Parsimony tree (B), the ML tree (C) and the BI tree (D), all rooted on the Gram positive organisms. In Dendroscope, open the `All.tre` file to display the four trees tiled in a single window. While holding down the **Shift** key click on each of the trees to select it (the background will become pink).

Figure 15.17

(A) NJ

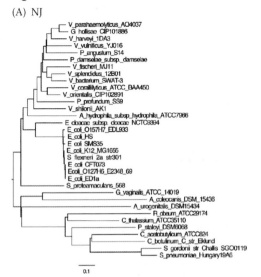

(B) Parsimony

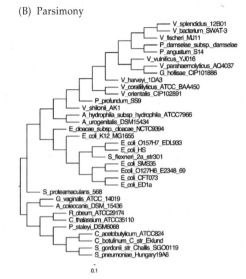

(C) ML

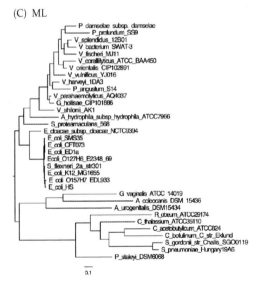

(D) BI

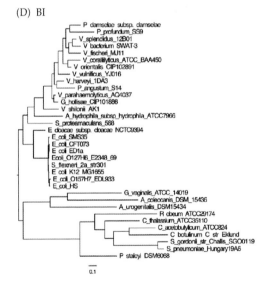

From the **Algorithms** menu choose **Network Consensus....** An **input** window opens that allows you to enter a threshold for inclusion of clusters in the network. This works exactly as described above for SplitsTree: the value you choose sets the minimum fraction of trees that must contain a cluster for that cluster to be included in the network. Finally, a window appears that allows you to choose the network algorithm (**Figure 15.18**).

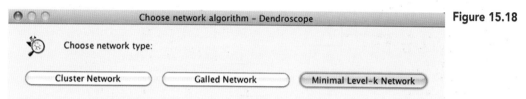

Figure 15.18

The **Cluster Network** option produces a hardwired network by the cluster-popping algorithm. The **Minimal Level-k Network** and **Galled Network** options produce softwired networks by the Minimal Level-k and Galled algorithms, respectively (see *Learn More about Phylogenetic Networks*, pp. 227–228). **Figure 15.19** shows the Cluster network (A), the Galled network (B), and Minimal-k networks (C) and (D) from the four trees in Figure 15.17.

Figure 15.19

(A)

(B)

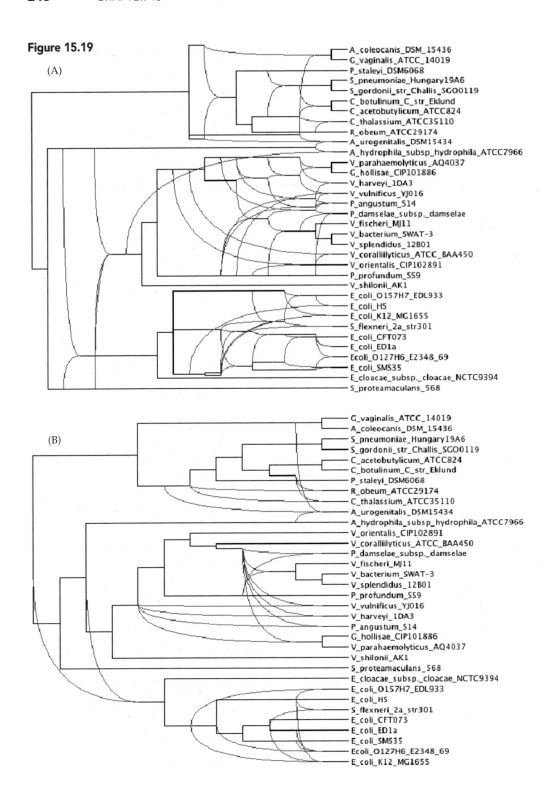

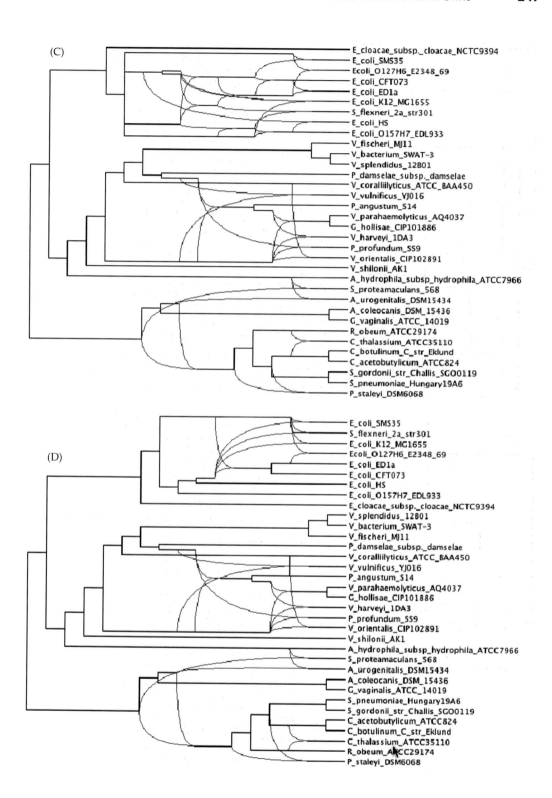

Estimating a minimal-k network can take awhile, so a progress bar (**Figure 15.20**) is displayed during the estimation. If the process appears to get hung up while processing a component, you can click the **Skip** button to allow things to continue. Figure 15.19C shows the network that resulted from skipping the processing of component 3 in the minimal-k network.

Figure 15.20

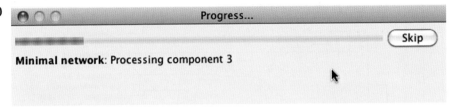

The drawings in Figure 15.19 may appear more difficult to interpret than the slanted format shown in Figure 4 of *Learn More about Phylogenetic Trees* (see Chapter 6, p. 74), but Dendroscope warns us that the slanted format is not well defined for networks and that the drawing may contain errors. Figure 15.19 demonstrates that the softwired networks are indeed less complex than the hardwired network. It appears that the minimal-k network is the default (see Figure 15.18) for a reason, and it also appears that it is worth the time required to process all components (compare Figures 15.19C and D).

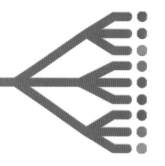

CHAPTER **16**

Some Final Advice: Learn to Program

Computer programming has become an essential skill for biologists at every level and in every field. We are confronted on almost a daily basis with the need to use output files generated either by computer programs or by instruments such as automated DNA sequencers. Often it is necessary to move information from such files into a specialized program for further analysis. One way to do that, of course, is to cut-and-paste the necessary information between programs. With increasingly large data sets and the consequently large output files, however, that can become an overwhelming task and is subject to increasing human error as fatigue sets in.

The solution is often to try to find a computer programmer to write a program to solve the repetitive task for you. That means finding the appropriate person, explaining what needs to be done, and then working with the result until things are right. If that sounds familiar, it is because that is probably what you used to do when you needed a phylogenetic analysis. If you have completed this book, you have freed yourself from that dependency on others and in the process have discovered that doing phylogenetic analysis is not as difficult as it once appeared.

It turns out that learning to write simple computer programs is no more difficult than learning to do phylogenetic analysis and frees you in much the same manner. Once while discussing with a student her overwhelming problem of reformatting some files for downstream analysis, I suggested that she learn to program in Perl enough to write a script to do the job. She responded "Oh, I couldn't do that. I don't even understand binary numbers!" I assured her that understanding binary numbers is not a prerequisite to programming, and that the return on her time investment would be well worth it.

The two Utility programs described in this book (see Chapter 11) were written in response to my own needs. They are simple and required only the most basic skills. FastaConvert was written because, despite the variety of format conversion programs in existence, none met my need to write files in Nexus, standard PHYLIP, and the special PHYLIP format required by codeml. ExtAncSeqMEGA was written to allow me to use MEGA's output in a way the authors had not considered. A sequence-evolution simulation program was written because nobody thought about simulating evolution in the way that made the most sense to me. The point is that no one else was going to meet my needs; it was up to me to do the programming or do without.

I strongly urge you to learn to program, whatever the current stage of your career as a biologist. A new book, *Practical Computing for Biologists* (Haddock and Dunn 2011), approaches programming in much the same way that this book approaches phylogenetic analysis: it treats programming in terms of tasks that are relevant to biologists. Rather than dealing with the arcana of programming in some language, it approaches programming (in a language called Python) as just another step in learning to use your computer effectively for biological research. It is also a very handy reference, full of tips that make everyday computer use simpler, easier, and less frustrating.

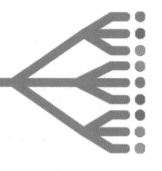

File Formats and Their Interconversion

One of the more frustrating aspects of using phylogenetics software is the plethora of file formats. MEGA uses its own format; MrBayes uses the Nexus format; PAML programs use a slightly unusual dialect of PHYLIP. If all you will ever do is create alignments using MEGA and then create phylogenetic trees from those alignments, also using MEGA, you don't need to know any more about file formats. If, however, you want to use other software for phylogenetic analyses, you need to know more about formats.

Format Descriptions

The MEGA format

The MEGA format (**Figure A1.1**) is read directly by MEGA and has the extension `.meg`. It always begins with `#mega` on the first line, followed by a series of lines that constitute the header information. Each header line begins with an exclamation point and ends with a semicolon. The title line begins `!Title` followed on that line by a title for the file (optional) and ending with a semi-

```
#mega
!Title ;
!Format DataType=DNA indel=- CodeTable=Standard;

!Domain=Data property=Coding CodonStart=1;
#E_coli_K12_MG1655
ATGAGGATCATCGATAACTTAGAACAGTTCCGCCAGATTTACGCCTCTGGCAAGAAGTGG
CAACGCTGCGTTGAAGCGATTGAAAATATC------GACAACATTCAGCCTGGCGTCGCC
CACTCCATCGGTGACTCATTGACTTACCGCGTGGAGACA------GACTCCGCG---ACC
GATGCGCTATTTACCGGGCATCGACGCTATTTTGAAGTGCATTACTACCTGCAAGGGCAG
CAAAAAATTGAATATGCGCCGAAAGAGACATTACAGGTAGTGGAATATTATCGTGATGAA
ACTGACCGTGAATATTTAAAA---GGCTGC---GGAGAAACCGTTGAGGTCCACGAAGGG
CAAATCGTTATTTGCGATATCCATGAAGCGTATCGGTTTATCTGCAAT------------
--------------AACGCGGTCAAAAAAGTGGTTCTCAAAGTCACCATCGAAGATGGT
---TATTTCCATAACAAA
|
#S_flexneri_2a_str301
ATGAGGATCATCGATAACTTAGAACAGTTCCGCCAGATTTACGCCTCTGGCAAGAAGTGG
CAACGCTGCGTTGAAGCGATTGAAAATATC------GACAACATTCAGCCTGGCGTCGCC
CACTCCATCGGTGATTCATTGACCTACCGTGTGGAAAAT------GACTCCGCG---ACC
GATGCGCTATTTACCGGGCATCGACGCTATTTTGAAGTGCATTACTACCTGCAAGGGCAG
CAAAAAATTGAATATGCGCCGAAAGAGACATTACAGGTAGTGGAATATTATCGTGATGAA
ACTGACCGTGAATATTTAAAA---GGCTGC---GGAGAAACCGTAGAGGGTCCACGAAGGG
CAAATCGTTATTTGCGATATCCATGAAGCGTATCGGTTTATCTGCAAT------------
--------------AACGCGGTCAAAAAAGTGGTTCTCAAAGTCACCATCGAAGATGGT
---TATTTCCATAACAAA
```

Figure A1.1 MEGA format

colon. The format line begins !`Format` followed by a series of options such as `DataType = DNA indel = - CodeTable = Standard` and ends with a semicolon. If the sequence is a coding sequence, the next line is !`Domain=Data property = Coding CodonStart = 1;`.

The header lines are followed by the sequences, consisting of an identifier line and then the sequence itself. The identifier line begins with a # symbol followed by the sequence name. There is no space between # and the name, and the name must not include spaces. Underscores in names are replaced by spaces in the tree drawings. The name may be followed by a comment enclosed in double quotes. The sequence begins on the next line and may be broken up into separate lines for better readability. Everything until the next # symbol is part of the sequence.

The FASTA format

In the FASTA format (**Figure A1.2**), each sequence is preceded by an identification line that starts with the symbol >. Everything on that line is treated as identification, not as part of the sequence. The line is terminated by a "hard" line break or return character. On the identification line, everything up to the first space is treated as the sequence name; everything after the space is treated as a comment and ignored.

Figure A1.2   FASTA format

```
>E_coli_K12_MG1655
ATGAGGATCATCGATAACTTAGAACAGTTCCGCCAGATTTACGCCTCTGGCAAGAAGTGG
CAACGCTGCGTTGAAGCGATTGAAAATATC------GACAACATTCAGCCTGGCGTCGCC
CACTCCATCGGTGACTCATTGACTTACCGCGTGGAGACA------GACTCCGCG---ACC
GATGCGCTATTTACCGGGCATCGACGCTATTTTGAAGTGCATTACTACCTGCAAGGGCAG
CAAAAAATTGAATATGCGCCGAAAGAGACATTACAGGTAGTGGAATATTATCGTGATGAA
ACTGACCGTGAATATTTAAAA---GGCTGC---GGAGAAACCGTTGAGGTCCACGAAGGG
CAAATCGTTATTTGCGATATCCATGAAGCGTATCGGTTTATCTGCAAT------------
---------------AACGCGGTCAAAAAAGTGGTTCTCAAAGTCACCATCGAAGATGGT
---TATTTCCATAACAAA
>S_flexneri_2a_str301
ATGAGGATCATCGATAACTTAGAACAGTTCCGCCAGATTTACGCCTCTGGCAAGAAGTGG
CAACGCTGCGTTGAAGCGATTGAAAATATC------GACAACATTCAGCCTGGCGTCGCC
CACTCCATCGGTGATTCATTGACCTACCGTGTGGAAAAT------GACTCCGCG---ACC
GATGCGCTATTTACCGGGCATCGACGCTATTTTGAAGTGCATTACTACCTGCAAGGGCAG
CAAAAAATTGAATATGCGCCGAAAGAGACATTACAGGTAGTGGAATATTATCGTGATGAA
ACTGACCGTGAATATTTAAAA---GGCTGC---GGAGAAACCGTAGAGGTCCACGAAGGG
CAAATCGTTATTTGCGATATCCATGAAGCGTATCGGTTTATCTGCAAT------------
---------------AACGCGGTCAAAAAAGTGGTTCTCAAAGTCACCATCGAAGATGGT
---TATTTCCATAACAAA
```

The sequence, without any numbering, follows. The sequence may consist of one long line without any line breaks, or it may include line breaks. Everything will be treated as part of the sequence until the next > character is encountered. Most programs ignore spaces.

FASTA is both the simplest and the most common sequence file format. Virtually all programs will recognize the FASTA format.

The Nexus format

The Nexus format is read directly by MrBayes and many other programs. It always begins with **#NEXUS**. The Nexus format allows the use of various "blocks," each of which starts with `Begin <block name>;` and ends with `end;`. The semicolons are essential elements in the Nexus format, indicating the end of a statement.

There are two versions of the Nexus format, often referred to as the "early" or "PAUP3" and the "late" or "PAUP4" format. Within each version there is an **interleaved** and a **sequential** format.

PAUP3 AND PAUP4 NEXUS FORMATS   In the PAUP3 format, the first block must be the data block. The alignment is part of the data block, which starts `Be-gin data;` (**Figure A1.3**). The next line is the `Dimensions` statement in which **ntax** is the number of sequences and **nchar** is the number of characters in the alignment. The `Format` statement on the following line must begin with the **datatype**, which says whether the alignment is of protein, DNA, or RNA sequences. It may list the symbols that are used in the alignment, and it should specify the symbol that is used for a gap. Again, each statement ends with a semicolon. Indentation is not necessary; it appears in Figure A1.3 only to make the file more readable to us.

The word `Matrix` indicates that what follows is the alignment.

```
Begin data;
    Dimensions ntax=34 nchar=498;
    Format datatype= DNA gap=-;
    Matrix
E_coli_K12_MG1655 ATGAGGATCATCGATAACTTAGAACAGTTCCGCCAGATTTACGC(
S_flexneri_2a_str301 ATGAGGATCATCGATAACTTAGAACAGTTCCGCCAGATTTA(
E_coli_O157H7_EDL933 ATGAGGATCATCGATAACTTAGAACAGTTCCGCCAGATTTA(
E_cloacae_subsp._cloacae_NCTC9394 ATGATCGTTCTTGAGAGCCTGGAACAGT1
P_damselae_subsp._damselae ATGATCGTACTTGATAACTTAGAGCAATTTAAATC(
V_harveyi_1DA3 ATGATCGTGTTAGACAACTTAGAACAATTCAAAGTCGTGTACCGCGA(
V_shilonii_AK1 ATGATCATCTTAGAGAGCTTAGAACAATTTAAGTCGGTCTATCGTAA1
V_vulnificus_YJ016 ATGATCGTGTTAGACAGCTTAGAGCAATTCAAACAGGTATATC(
V_coralliilyticus_ATCC_BAA450 ATGATCGTGTTAGAAAACTTAGCGCAATTCAA/
V_parahaemolyticus_AQ4037 ATGATCGTGTTAGACAGCTTAGAGCAATTCAAACAG(
V_splendidus_12B01 ATGATCGTTTTAGACAACCTAGAGCAATTTAAAGTCGTTTACC(
V_orientalis_CIP102891 ATGATCGTGTTAGAGAGCTTAGAGCAATTCAAAGTGGTT1
P_profundum_SS9 ATGGTTGTTTTAGAGAACCTCGAGCAATTCAAAACATTATACCGCG/
G_hollisae_CIP101886 ATGATCGTGTTAGACAGCTTGGCGCAATTCAAACAGGTATA1
P_angustum_S14 ATGATCGTATTAGATAGCTTAGCGCAATTTAAATTGGTGTACCGCAA1
V_bacterium_SWAT-3 ATGATCGTTTTAGAGAACTTAGAACAATTTAAAGTCGTCTACC(
```

Figure A1.3   Nexus PAUP3 sequential format

The PAUP4 version of Nexus involves a **taxa block** that lists the names of all the sequences, followed by a **characters block** that includes the matrix of sequences (**Figure A1.4**).

```
#NEXUS

begin taxa;
    dimensions ntax= 34;
    taxlabels
        E_coli_K12_MG1655
        S_flexneri_2a_str301
        E_coli_O157H7_EDL933
        .

        .

        Ecoli_O127H6_E2348_69
        E_coli_ED1a
        E_coli_HS
;
end;
begin characters;
    dimensions nchar= 498;
    format missing=? gap=- matchchar=. datatype=DNA;

    matrix

E_coli_K12_MG1655 ATGAGGATCATCGATAACTTAGAACAGTTCCGCCAGATTTACGCCTCTGGCAAGAAGTGGCAACGCTG
S_flexneri_2a_str301 ATGAGGATCATCGATAACTTAGAACAGTTCCGCCAGATTTACGCCTCTGGCAAGAAGTGGCAACG
E_coli_O157H7_EDL933 ATGAGGATCATCGATAACTTAGAACAGTTCCGCCAGATTTACGCCTCTGGTAAGAAGTGGCAACG
.

.

Ecoli_O127H6_E2348_69 ATGAGGATCATCGATAACTTAGAACAGTTCCGCCAGATTTACGCCTCTGGCAAGAAGTGGCAAC
E_coli_ED1a ATGAGGATCATCGATAACTTAGAACAGTTCCGCCAGATTTATGCCTCTGGCAAGAAGTGGCAACGATGCGTTGA
E_coli_HS ATGAGGATCATCGATAACTTAGAACAGTTCCGCCAGATTTACGCCTCTGGTAAGAAGTGGCAACGCTGCGTTGAAG
;
end;
```

Figure A1.4   Nexus PAUP4 sequential format

SEQUENTIAL AND INTERLEAVED FORMATS   Nexus provides for two data formats, sequential and interleaved. In the sequential format, each sequence begins on a new line with the sequence name, followed by at least one space (see Figures A1.3 and A1.4). What comes after that is the aligned sequence. In the two figures above, the sequences extend out to the right past the edge of the figure.

In the interleaved format (**Figure A1.5**) each taxon name, followed by a fixed number of characters, is written on a single line; a blank line is written, and the process is repeated. Note that the format statement includes the word `inter-leave`. Without that information, programs are unable to interpret the interleaved format. Some programs, such as MrBayes, require `interleave=yes`, which is more formally correct in the Nexus format; others are satisfied with just the word `interleave`. Most, but not all, programs that use the Nexus format will handle either sequential or interleaved data.

In Figure A1.5, most of the sequences are replaced by a dot so that you can see the repeating sections of sequence.

```
begin data;
    dimensions ntax=34 nchar=498;
    format missing=? gap=- matchchar=. datatype=DNA interleave=yes;
    matrix

E_coli_K12_MG1655          ATGAGGATCATCGATAACTTAGAACAGTTCCGCCAGATTTACGCCTCTGGCAAGAAGTGGCAACGCTGCGTTGAAGCG
S_flexneri_2a_str301       ATGAGGATCATCGATAACTTAGAACAGTTCCGCCAGATTTACGCCTCTGGCAAGAAGTGGCAACGCTGCGTTGAAGCG
.
.
Ecoli_O127H6_E2348_69      ATGAGGATCATCGATAACTTAGAACAGTTCCGCCAGATTTACGCCTCTGGCAAGAAGTGGCAACGATGCGTTGAAGCG
E_coli_ED1a                ATGAGGATCATCGATAACTTAGAACAGTTCCGCCAGATTTATGCCTCTGGCAAGAAGTGGCAACGATGCGTTGAAGCG
E_coli_HS                  ATGAGGATCATCGATAACTTAGAACAGTTCCGCCAGATTTACGCCTCTGGCAAGAAGTGGCAACGCTGCGTTGAAGCG

E_coli_K12_MG1655          ATTGAAAATATC------GACAACATTCAGCCTGGCGTCGCCCACTCCATCGGTGACTCATTGACTTACCGCGTGGAG
S_flexneri_2a_str301       ATTGAAAATATC------GACAACATTCAGCCTGGCGTCGCCCACTCCATCGGTGATTCATTGACCTACCGTGTGGAA
E_coli_O157H7_EDL933       ATTGAAAATATC------GACAACATTCAGCCTGGCGTCGCCCACTCCATCGGTGACTCATTGACTTACCGCGTGGAG
.
.
Ecoli_O127H6_E2348_69      ATTGAAAATATC------GACAACATTCAGCCTGGCGTCGCCCACTCCATCGGTGACTCATTGACTTACCGCGTGGAG
E_coli_ED1a                ATTGAAAATATC------GACAACATTCAGCCTGGCGTCGCCCACTCCATCGGTGACTCATTGACTTACCGCGTGGAG
E_coli_HS                  ATTGAAAATATC------GACAACATTCAGCCTGGCGTCGCCCACTCCATCGGTGACTCATTGACTTACCGCGTGGAG

E_coli_K12_MG1655          ACA------GACTCCGCG---ACCGATGCGCTATTTACCGGGCATCGACGCTATTTTGAAGTGCATTACTACCTGCAA
S_flexneri_2a_str301       AAT------GACTCCGCG---ACCGATGCGCTATTTACCGGGCATCGACGCTATTTTGAAGTGCATTACTACCTGCAA
```

Figure A1.5 Nexus PAUP3 interleaved format

COMMENTS IN NEXUS The Nexus format allows the user to "comment out" statements by enclosing them in square brackets, [ ]. Statements surrounded by square brackets are ignored when the file is executed, which can be very convenient when you want to try something out and do not want all of the instructions to be executed. For instance, during a preliminary runs with MrBayes, you may only be interested in the time required per generation, so you may not want to have the various files written by the `sumt` command. By enclosing the `sumt` statement in square brackets so that it reads [`sumt relburnin = yes burnin frac = 0.25 contype = allcompat;`], the entire statement will be ignored.

Note that the square brackets can comment out more than one statement at a time, and they need not be on the same line. Everything between [ ] is ignored by programs that use the Nexus format.

TREE DESCRIPTIONS IN NEXUS The Nexus format provides for tree descriptions in the form of "trees blocks." **Figure A1.6** shows a Nexus file that describes two trees, each with an explanation provided in the form of a comment preceding the tree. Each tree description begins with the key word `tree` followed by the tree name, an equals sign, and the Newick tree description.

```
#NEXUS

begin trees;
    [Note: This tree contains information on the topology,
          branch lengths (if present), and the probability
          of the partition indicated by the branch.]
    tree con_50_majrule = (E_coli_K12:0.002888,E_coli_E24377A:0.005495,E_coli_UTI89:0.00328

|   [Note: This tree contains information only on the topology
          and branch lengths (mean of the posterior probability density).]
    tree con_50_majrule = (E_coli_K12:0.002888,E_coli_E24377A:0.005495,E_coli_UTI89:0.00328
end;
```

Figure A1.6 Nexus PAUP3 tree file

The tree names need not be unique. In figure A1.6, the Newick description extends beyond the right edge of the figure. The tree description can be copied and pasted into another file to make a simple Newick (PHYLIP) tree file.

The PAUP4 version of a tree description uses a taxa block and may incorporate a translation table (**Figure A1.7**). Within the translation table, each taxon is numbered and there is a comma following the taxon name. In the Newick tree descriptions, taxon names are replaced by the corresponding number from the translation table. As a result the tree description cannot simply be copied and used elsewhere.

```
#NEXUS

[ID: 8964260026]
begin taxa;
    dimensions ntax=34;
    taxlabels
        E_coli_K12_MG1655
        S_flexneri_2a_str301
        E_coli_O157H7_EDL933
        .
        .
        .
        Ecoli_O127H6_E2348_69
        E_coli_ED1a
        E_coli_HS
        ;end;
begin trees;
    translate
        1    E_coli_K12_MG1655,
        2    S_flexneri_2a_str301,
        3    E_coli_O157H7_EDL933,
        .
        .
        .
        32   Ecoli_O127H6_E2348_69,
        33   E_coli_ED1a,
        34   E_coli_HS
    ;
    [Note: This tree contains information on the topology,
           branch lengths (if present), and the probability
           of the partition indicated by the branch.]
    tree con_50_majrule = (1:0.003779,2:0.007976,(4:0.175458,((((((((5:0.169894,(((9:0.17

    [Note: This tree contains information only on the topology
           and branch lengths (mean of the posterior probability density).]
    tree con_50_majrule = (1:0.003779,2:0.007976,(4:0.175458,(((((((((5:0.169894,(((9:0.17
end;
```

Figure A1.7   Nexus PAUP4 tree file with translation table

The PHYLIP format

PHYLIP (Phylogenetic Inference Package) is a very popular (and free) set of programs for phylogenetic analysis and the creation of phylogenetic trees. Because of the program's popularity and the simplicity of its format, many other programs, including the PAML suite that includes codeml, use the PHYLIP format.

There are several versions of the PHYLIP format. In the original version, taxon names had to be exactly 10 characters; shorter names were "padded" with spaces. The extended version allows names up to 50 characters and separates the name from the sequence by *one* space. PAML allows up to 30 characters and separates the name from the sequence by *two* spaces.

Whichever version is being used, a PHYLIP file always begins with a header line that that gives the number of sequences followed by the number of characters in each sequence. PHYLIP permits the sequences to be interleaved, but if they are interleaved the letter I must follow the number of characters on the header line.

```
|12 273
U68496  GTAGTAATTAGATCTGAAAACTTCTCGAACAATGCTAAAACCATAATAGTACAGCT.
U68497  ATAGTAATTAGATCTGAAAACTTCTCGAACAATGCTAAAACCATAATAGTACAGCT.
U68498  GTAGTAATTAGATCTGAAAACTTCACGAACAATGCTAAAACCATAATAGTACAGCT.
U68499  ATAGTAATTAGATCTGAAAACTTCACGAACAATGCTAAAACCATAATAGTACAGCT.
U68500  ATAGTAATTAGATCTGAAAACTTCACGAACAATGCTAAAACCATAATAGTACATCT.
U68501  GTAGTAATTAGATCTGAAAACTTCACGAACAATGCTAAGACCATAATAGTACAGCT.
U68502  GTAGTAATTAGATCTGAAAACTTCACGAACAATGCTAAGACCATAATAGTACAGCT.
U68503  ATAGTAATTAGATCTGAAAACTTCACGAACAATGCTAAAACCATAATAGTACAGCT.
U68504  ATAGTAATTAGATCTGAAAACTTCACAGACAATGCTAAAACCATAATAGTACAGCT.
U68505  ATAGTAATTAGATCTGAAAACTTCACAGACAATGCTAAAACCATAATAGTACAGCT.
U68506  ATAGTAATTAGATCTGAAAACTTCACGGACAATGCTAAAACCATAATAGTACAGCT.
U68507  GTAGTAATTCGATCTGAAAACTTCACGGACAATGCTAAAACCATAATAGTACAGCT.
```

Figure A1.8 PHYLIP sequential format, PAML variant

Interconverting Formats

FastaConvert and MEGA

As described in this book, you can convert the FASTA format to the sequential Nexus, PHYLIP extended, and PHYLIP PAML formats with the FastaConvert utility program (see Chapter 11).

Alternatively, MEGA can export files in the MEGA format (.meg) to the PHYLIP format and to PAUP3 and PAUP4 versions of the Nexus format. Open the .meg file in MEGA's main window and click the large **TA** icon to open the **Sequence Data Explorer**. In the Sequence Data Explorer window, click **TA** so that all of the characters are shown, then from the **Data** menu choose **Export Data**. In the resulting dialog choose the format you want, tick the **Interleaved Output** if you want the interleaved format, then click **OK**.

The converted alignment will appear in the **Text File Editor and Format Converter** window. Save the file from that window.

Other format conversion programs

Two other conversion programs are noteworthy; both are free and are available for Windows and Mac.

SeaView (http://pbil.univ-lyon1.fr/software/seaview.html) reads and writes MASE, Phylip, Clustal, MSF, FASTA, and Nexus. SeaView also aligns sequences by MUSCLE and ClustalW and estimates trees by Parsimony, NJ, and ML (using PhyML). See Appendix II for more information.

DataConvert (http://biology.byu.edu/faculty/dam83/cdm/) reads and writes PHYLIP, Clustal, MEGA, Nexus (PAUP3), and FASTA formats in both sequential and interleaved formats. The PHYLIP format truncates taxon names to 10 characters. DataConvert has the interesting feature that it permits selecting which sequences are to be written, which is useful if (for instance) you need to eliminate specific sequences from an interleaved file.

Additional Programs

The programs discussed in this book are intended to provide a basic set of tools for estimating and presentating phylogenetic trees. I have not attempted to provide comprehensive coverage of the hundreds of phylogenetics programs that are available, and you may encounter situations in which none of the programs described here does exactly what you want. At that point you need to start exploring the variety of other programs that exist.

Some programs are easy to use, some are difficult. Some run on several computer platforms, others only on one. Some are only available as source code, requiring you to compile them for your computer. Some programs that are available for Mac and Mac OSX will not run on the new Intel-based Macintoshes; Intel-Mac users will need to download the source code and compile it.

Inclusion of a program here should not be taken as an endorsement or assurance that the program will perform as described, nor should the absence of a program from this list be taken as a condemnation. My choice of which programs to include is based somewhat on personal experience, somewhat on the general reputation of the program, and somewhat on the suggestions of colleagues. My list is far from complete, but Joseph Felsenstein of the University of Washington has compiled and maintains a very comprehensive list of phylogenetics programs that can be seen at http://evolution.genetics.washington.edu/phylip/software.html. It is superbly organized, categorized, and cross-referenced according to virtually any criterion you could imagine. If you don't see what you want here, check out Joe's list—if a program exists, it is almost certainly there.

BALI-PHY Jointly estimates Bayesian trees and alignment. Unlike most programs that treat the sequence alignment as fixed (and thus, in effect, as "true"), Bali-Phy uses a Bayesian approach to consider multiple near-perfect alignments as it considers the set of near-perfect trees. **Platforms:** Mac, Windows, and as source code. Download from http://www.biomath.ucla.edu/~msuchard/bali-phy/

BAMBE Bayesian Inference from DNA sequence data. **Platforms:** Windows, Unix. Download from http://www.mathcs.duq.edu/larget/bambe.html.

BAMBE can also be run from a web server: http://bioweb.pasteur.fr/seqanal /phylogeny/intro-uk.html

GENEIOUS Multifunctional program that downloads and manages not only nucleic acid and protein sequences, but references as well. Aligns sequences, makes NJ trees, and (with free plug-ins) runs MrBayes and PHYML to make trees by BI and ML. Free for academic use, or purchase the Professional version with even more features. **Platforms:** Mac, Windows, and Linux. http:// www.geneious.com/

JMODELTEST Not a phylogenetics program, but a program that tests the hierarchies of various models using the likelihoods of ML trees under those models. **Platforms:** Mac, Windows, Linux. http://darwin.uvigo.es/software/ jmodeltest.html

MAC5 Estimates trees by Bayesian Inference, but differs from other programs in that it can consider gaps as information rather than just as missing data. **Platforms:** Windows, Mac, and as source code (Unix). Intel-Mac users will need to compile the Unix source code themselves. http://www.agapow.net/ software/mac5

MACCLADE Not a phylogenetics program *per se*, but an interactive tree manipulation program that allows the user to rearrange phylogenies by hand. It can also edit data and simulate the evolution of data on a tree. **Platforms:** Macintosh only. http://macclade.org/macclade.html

PAML Powerful but not particularly easy to use, PAML includes many functions beyond the codeml described in Chapter 14. Estimates trees by ML and BI, tests evolutionary models, and so forth. **Platforms:** Windows, Mac, and as source code. http://abacus.gene.ucl.ac.uk/software/paml.html

PAUP* A multifeatured program that estimates trees by all methods except BI. Although it estimates ML trees from DNA data, it cannot do so from protein sequence data. PAUP* outperforms MEGA in estimating trees by parsimony, particularly in calculating branch lengths. The Macintosh classic (pre-OSX) version, which employs the typical Macintosh interface with pull-down menus, dialog boxes, etc., includes an excellent tree-drawing module, but that version cannot run on Intel-Macs. The versions for Windows, Linux, several Unix machines, Mac OSX, and Intel-Macs are command-line versions that require entering the various commands as part of a PAUP block in the Nexus input file (see Appendix I).

Although it is intimidating for the inexperienced, PAUP* is one of the most popular phylogenetics programs because it was one of the first to provide so many functions within a single program. First released in 1999, the current version (released in 2003) is still a beta-release (v4.0b10). No manual is available, but a Command Reference document is provided. Detailed instructions for using PAUP* are provided in the second edition of *Phylogenetic Trees Made Easy* (Hall 2004). PAUP* is available for purchase (prices vary depending upon platform) from Sinauer Associates http://www.sinauer.com/detail. php?id=8060. Also see the PAUP* main page at http://paup.csit. fsu.edu/

PHASE   Especially designed for estimating trees by Bayesian Inference from RNA sequences, its models include both paired sites and unpaired sites. **Platforms:** Windows, Linux, and as C++ source code. http://www.bioinf.man. ac.uk/resources/phase/

PHYLIP   The granddaddy of all phylogenetics packages, distributed since 1980 and with over 20,000 registered users. It estimates trees by NJ and other distance methods, by Parsimony, and by Maximum Likelihood. PHYLIP is not a single program but a suite (a very large suite, more a mansion) of nearly 50 individual programs that share a common interface. Because each individual task requires invoking a different program, PHYLIP is less convenient to use than MEGA, but it is very versatile. **Platforms:** Windows, Mac, and as source code from http://evolution.gs.washington.edu/phylip.html (Don't miss the link to "Credits for P" on that page, including the "No thanks to…" section.)

PHYML   A fast program for estimating ML trees, its use was discussed in detail in the third edition of *Phylogenetic Trees Made Easy* (Hall 2008). PhyML provides the aLRT algorithm for calculating branch (clade) support. Unlike the bootstrap algorithm, aLRT requires no additional time. For ease of use you may prefer to run PhyML through SeaView (see below). **Platforms:** Windows, Mac, Linux. PhyML can also be run online over the web at http://atgc.lirmm. fr/phyml/. The stand-alone program may be downloaded from http://www. atgc-montpellier.fr/phyml/binaries.php

PROTTEST   Tests 64 different models of protein evolution and uses PHYML to estimate likelihoods. Its purpose is similar to that of jModelTest described above. **Platforms:** Mac, Windows, and Linux. Download from http://darwin. uvigo.es/software/prottest.html

SEAVIEW   A multi-purpose program, SeaView aligns sequences by ClustalW and MUSCLE; estimates trees by NJ, Parsimony and ML. SeaView uses PhyML to estimate ML trees and may be more convenient to use than the original stand-alone program. (See the description of PhyML above for some reasons to use PhyML.) It is also an excellent means of converting among various sequence alignment formats.

Frequently Asked Questions

1. What if I want to use MEGA to download the sequences, but want to use another program for aligning them?

You need to *export* the sequences from the Alignment Explorer and *save* them in a format that the other alignment program can use. The first step is to look at the documentation for your chosen alignment program to determine what format(s) it can use. Next, read Appendix I to learn more about file formats.

From the **Data** menu of the **Alignment Explorer**, choose **Export Alignment** (**Figure A3.1**). You are offered three choices: **MEGA format**, **FASTA format**, and **PAUP format**. The other program is unlikely to accept MEGA format (but it is possible that it will). FASTA is probably the most widely used format. What MEGA calls the "PAUP format" is more properly called the Nexus format. It is highly likely that your alignment program will accept one of these two formats, but if not you may be able to export the sequences in FASTA or PAUP (Nexus) format, then use another program to convert that format to one that is recognized by the other program. (See Appendix I for tips on interconverting formats.)

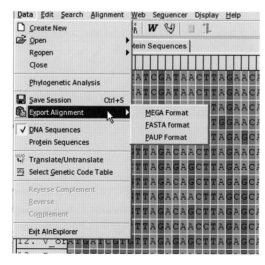

Figure A3.1

2. What if I already have a nucleotide alignment created by another program and I want to use MEGA to align those sequences by codons?

You can import an existing alignment directly into the alignment editor. From MEGA's main window choose **Edit/Build Alignment** from the **Align** menu. In the resulting window (Figure A3.2), choose **Retrieve sequence from file** and navigate to the folder where the existing alignment file is located. MEGA can import files in FASTA, Nexus (PAUP), and Clustal formats. In order to recognize those files, they must have one of the extensions shown in Figure A3.2. Select the file and open it. The alignment will appear in a new **Alignment Explorer** window.

Figure A3.2

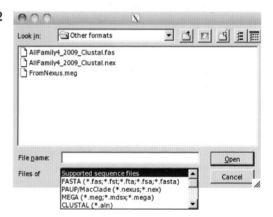

If an Alignment Explorer window is already open you can accomplish the same thing by choosing **Open** and then **Retrieve sequence from file** from the Alignment Explorer's **Data** menu. The imported alignment will replace the one currently in the Alignment Explorer. Once you have imported the alignment, type `Control+A` to select all the sequences, then choose **Delete Gaps** from the **Edit** menu. You can now align those sequences by codons using either MUSCLE or ClustalW.

3. Why should I always save alignments as .mas files?

It may seem wasteful to save files in the .mas format when the alignment editor can retrieve sequences from a .meg file. However, the .meg format converts all the underscores in the sequence names to spaces, and the sequence names are shown with spaces on the tree drawings. That certainly looks better, but many programs will not accept sequences names with spaces. If you need to export an alignment for use by another program, you need to export it from the Alignment Editor with those underscores intact. If the Alignment Editor retrieves a sequence from a .meg files the spaces will be missing. Since it may be difficult to anticipate future needs, *always* save your alignment in the .mas format by choosing **Save Session** from the Alignment Editor's **Data** menu.

4. What if I want to add sequences to an existing alignment?

If the alignment was saved by MEGA as a .mas file, just open that file by choosing **Open Saved Alignment File** from the **Align** menu of MEGA's main window. If it was saved as a .meg file, choose **Edit/Build Alignment** from the **Align** menu of MEGA's main window, then choose **Retrieve sequences from file**. **Delete Gaps** as described in Question 2, then add new sequences as described in Chapter 3. If the alignment is in a different format, import it as described in Question 2, **Delete Gaps**, then add the new sequences.

5. What if I want to analyze an alignment using a different phylogenetics program?

Export the alignment in either FASTA or Nexus format as described in Question 1.

6. What if I want to use MEGA to estimate a tree from an alignment created by another program?

Import the alignment as described in Question 2, then save it as a .meg file and estimate the tree.

7. What if I want to set a more stringent criterion than the majority rule for collapsing low-support branches to polytomies, perhaps 70%?

In the dialog shown in Figure 6.12 (p. 87), just change the cut-off value to whatever you choose.

8. Why don't you describe how to use PAUP*?

At one time PAUP* (see p. 260) was a very major program for phylogenetic analysis. It was discussed extensively in the first and second editions of *Phylogenetic Trees Made Easy* (Hall 2001, 2004). PAUP* has not been updated in over 7 years, however, and a promised manual has never been written. If you want a book to explain how to use PAUP*, I'm afraid you'll have to dig up a copy of the first or second edition.

9. Why don't you discuss estimating ancestral sequences using MrBayes?

Using MrBayes to estimate ancestral sequences, as first described in Hall 2006, is effective but is both time-consuming and labor-intensive. Estimation of the ancestral sequences requires two MrBayes runs (typically a few hours each) *for each node*, followed by processing those results with four separate Perl scripts. In contrast, MEGA generates the necessary output for all nodes simultaneously, typically in less than 3 minutes, followed by processing that output with a single Perl script that requires under a 1 minute. To obtain the ancestral sequences for all 62 internal nodes of a 64-taxon tree using MrBayes would require about 66 days of computer time, to say nothing of the hours of user time required to set up all of the input files and then process the output. MEGA does the same job in less than 5 minutes. Since ML estimation of ancestral sequences with MEGA is so much faster and more efficient and is just as accurate, there is little reason to discuss using MrBayes. If you prefer to use MrBayes, see the third edition of this book (Hall 2008)—but you will need to modify the input files appropriately to use MrBayes v3.2.

Literature Cited

Edgar, R. C. 2004a. MUSCLE: A multiple sequence alignment method with reduced time and space complexity. *BMC Bioinformatics* 5: 113.

Edgar, R. C. 2004b. MUSCLE: Multiple sequence alignment with high accuracy and high throughput. *Nucleic Acids Res.* 32: 1792–1797.

Felsenstein, J. 2004. *Inferring Phylogenies.* Sinauer Associates, Sunderland, MA.

Fletcher, W. and Z. Yang. 2010. The effect of insertions, deletions, and alignment errors on the Branch-Site test of positive selection. *Mol. Biol. Evol.* 27: 2257–2267.

Graur, D. and W.-H. Li. 2000. *Fundamentals of Molecular Evolution*, 2nd Ed. Sinauer Associates, Sunderland, MA.

Haddock, S. H. D. and C. W. Dunn. 2011. *Practical Computing for Biologists.* Sinauer Associates, Sunderland, MA.

Hall, B. G. 2001, 2004, 2008. *Phylogenetic Trees Made Easy*, 1st, 2nd, 3rd Eds. Sinauer Associates, Sunderland, MA

Hall, B. G. 2005. Comparison of the accuracies of several phylogenetic methods using protein and DNA sequences. *Mol. Biol. Evol.* 22: 792–802.

Hall, B. G. 2006. Simple and accurate estimation of ancestral protein sequences. *Proc. Natl. Acad. Sci. USA* 103: 5431–5436.

Hillis, D. M., C. Moritz and B. K. Mable. 1996. Applications of molecular systematics. In *Molecular Systematics,* pp. 515–543, D. M.

Hillis, C. Moritz and B. K. Mable (eds.). Sinauer Associates, Sunderland, MA.

Huson, D.H., R. Rupp and C. Scornavacca. 2011. *Phylogenetic Networks: Concepts, Algorithms and Applications.* Cambridge University Press, Cambridge.

Jukes, T. H. and C. R. Cantor. 1969. Evolution of protein molecules. In *Mammalian Protein Metabolism*, pp. 21–32, H. N. Munro (ed.). Academic Press, New York.

Kimura, M. 1980. A simple method for estimating evolutionary rates of base substitutions through comparative studies of nucleotide sequences. *J. Mol. Evol.* 16: 111–120.

Kumar, S. and A. Filipski. 2006. Multiple sequence alignment: In pursuit of homologous DNA positions. *Genome Res.* 17: 127–135.

Li, W.-H. 1997. *Molecular Evolution.* Sinauer Associates, Sunderland, MA.

Loytynoja, A. and N. Goldman. 2008. Phylogeny-aware gap placement prevents errors in sequence alignment and evolutionary analysis. *Science* 320: 1632–1635.

Mau, B. and M. Newton. 1997. Phylogenetic inference for binary data on dendrograms using Markov chain Monte Carlo. *J. Comp. Graph. Stat.* 6: 122–131.

Mau, B., M. Newton and B. Larget. 1999. Bayesian phylogenetic inference via Markov chain Monte Carlo methods. *Biometrics* 55: 1–12.

Nei, M. and S. Kumar. 2000. *Molecular Evolution and Phylogenetics*. Oxford University Press, New York.

Nuin, P. A., Z. Wang and E. R. Tillier. 2006. The accuracy of several multiple sequence alignment programs for proteins. *BMC Bioinformatics* 7: 471.

Ogden, T. H. and M. S. Rosenberg. 2006. Multiple sequence alignment accuracy and phylogenetic inference. *Syst. Biol.* 55: 314–328.

Penn, O., E. Privman, H. Ashkenazy, G. Landan, D. Graur et al. 2010. GUIDANCE: A web server for assessing alignment confidence scores. *Nucleic Acids Res.* 38(Suppl.): W23–W28.

Rannala, B. and Z. H. Yang. 1996. Probability distribution of molecular evolutionary trees: A new method of phylogenetic inference. *J. Mol. Evol.* 43: 304–311.

Ronquist, F. and J. P. Huelsenbeck. 2003. MrBayes 3: Bayesian phylogenetic inference under mixed models. *Bioinformatics* 19: 1572–1574.

Swofford, D. L., G. J. Olson, P. J. Waddell and D. M. Hillis. 1996. Phylogenetic inference. In *Molecular Systematics*, pp. 407–514, D. M. Hillis, C. Moritz and B. K. Mable (eds.). Sinauer Associates, Sunderland, MA.

Tamura, K., D. Peterson, N. Peterson, G. Stecher, M. Nei and S. Kumar. 2011. MEGA 5: Molecular Evolutionary Genetics Analysis using maximum likelihood, evolutionary distance, and maximum parsimony methods. *Mol. Biol. Evol.* (in press).

Tavaré, L. 1986. Some probabilistic and statistical problems on the analysis of DNA sequences. *Lect. Math. Life Sci.* 17: 57–86.

Thompson, J. D., D. G. Higgins and T. J. Gibson. 1994. ClustalW: Improving the sensitivity of progressive multiple sequence alignment through sequence weighting, position-specific gap penalties and weight matrix choice. *Nucleic Acids Res.* 22: 4673–4680.

Thompson, J. D., F. Plewniak and O. Poch. 1999. A comprehensive comparison of multiple sequence alignment programs. *Nucleic Acids Res.* 27: 2682–2690.

Yang, Z. 1997. PAML: A program package for phylogenetic analysis by maximum likelihood. *Comput. Appl. Biosci.* 13: 555–556.

Yang, Z., R. Nielsen, N. Goldman and A.-M. K. Pedersen. 2000. Codon-substitution models for heterogeneous selection pressure at amino acid sites. *Genetics* 155: 431–449.

Index to Major Program Discussions

Subject Index